Land Use and Climate Change Impacts on Erosion and Sediment Transport

Climate Change Impacts on Sediment Dynamics: Measurement, Modelling and Management

ICCE Symposium 2018

Proceedings of the International Association of Hydrological Sciences (PIAHS) Volume 381, 2019

Moscow, Russia
27-31 August 2018

Editors:

Sergey Chalov
Valentin Golosov
Adrian Collins
Mike Stone

ISBN: 978-1-5108-9544-7

Printed from e-media with permission by:

Curran Associates, Inc.
57 Morehouse Lane
Red Hook, NY 12571

Some format issues inherent in the e-media version may also appear in this print version.

International Association of Hydrological Sciences
Centre for Ecology and Hydrology
Wallingford, Oxfordshire OX10 8BB
United Kingdom

Phone: 441 491 692 442
Fax: 441 491 692 448

http://iahs.info/

Additional copies of this publication are available from:

Curran Associates, Inc.
57 Morehouse Lane
Red Hook, NY 12571 USA
Phone: 845-758-0400
Fax: 845-758-2633
Email: curran@proceedings.com
Web: www.proceedings.com

TABLE OF CONTENTS

Proc. IAHS, 381, 1–1, 2019
https://doi.org/10.5194/piahs-381-1-2019

Preface: Land use and climate change impacts on erosion and sediment transport

Sergey Chalov[1], **Valentin Golosov**[1,2,3], **Adrian Collins**[4], **and Mike Stone**[5]

[1]Faculty of Geography, Lomonosov Moscow State University, Moscow, Russia
[2]Kazan Federal University, Moscow, Russia
[3]Insitute of Geography RAS, Moscow, Russia
[4]Sustainable Agriculture Sciences Rothamsted Research, Okehampton, UK
[5]Faculty of Environment, University of Waterloo, Waterloo, Canada

Correspondence: Sergey Chalov (srchalov@geogr.msu.ru)

Published: 1 August 2019

The papers presented in this volume were part of a joint symposium combining The Second International Young Scientists Forum on Soil and Water Conservation (WASWAC) and the International Commission on Continental Erosion (ICCE-IAHS) which was held on 27–31 August 2018, in Moscow. Contributions were invited on work that investigates fluvial processes including hillslope and gully erosion and sediment transport, as well as river channel changes and management, the influence of hydro-sedimentological processes operating at local to global scales, and those that compare the findings drawn from local and global studies. A particular focus of the papers in this new volume is to provide regional examples of sediment movement along the sediment cascade in Northern Eurasia. Most of these works report scientific studies across a territory which covers nearly one sixth of the land surface of the Earth. PIAHS volume permits wider dissemination of these regional studies. Collectively, the papers cover studies from the East European plain, Kola Peninsula Mountains, Central Siberian Plateau, and the Kamchatka peninsula, as well as river catchments in Japan and Italy.

This PIAHS volume continues the long-term tradition of stand-alone proceedings from meetings convened by the International Commission on Continental Erosion (ICCE-IAHS). Since 1981, ICCE has arranged a sequence of conferences and published 36 proceedings – and this volume discusses the ICCE story through a detailed scientometric analysis of past ICCE activities (Bondarev, in this volume).

The first part of this volume comprises regional studies of soil and gully erosion (Sharifullin et al., in this volume; Gafurov et al., in this volume; Garankina et al., in this volume). The section also includes examples of an integrated approach to the study of sediment origin (Chalov et al., in this volume).

The second part of the volume is devoted to river process and connections with sediment transport issues. It includes novel studies of sediment yield presented by case studies from Italy (Porto and Callegari, in this volume), the Kamchatka peninsula (Kuksina, in this volume), Caucuses mountains (Tsyplenkov et al., in this volume) and channel erosion research (Golovlyov et al., in this volume; Fingert et al., in this volume; Kuznetsova et al., in this volume).

The third part of the volume links sediment transport processes and lake sedimentation, as well as environmental impacts. It mostly includes studies on the reconstruction of the long-term dynamics of sedimentation (Konoplev et al., in this volume; Wakiyama et al., in this volume; Garankina et al., in this volume), but also focuses on river runoff processes and salmon migration (Moreydo, in this volume).

The editors wish to thank all the contributing authors for their efforts and the numerous reviewers who contributed to the production of quality papers. Special acknowledgement is also extended to IAHS and WASWAC for facilitating the participation of young scientists. Additionally, this review is supported by RFBR project 18-05-60219.

Proc. IAHS, 381, 3–6, 2019
https://doi.org/10.5194/piahs-381-3-2019

Scientometric analysis of the activities of the International Commission for Continental Erosion (ICCE)

Valeriy P. Bondarev[1,2]

[1]Faculty of Geography, Lomonosov Moscow State University, Moscow, 119 991, Russia
[2]Faculty of Social Sciences and Humanities, Bauman Moscow State Technical University, Moscow, 105 005, Russia

Correspondence: Valeriy P. Bondarev (valery_bondarev@mail.ru)

Published: 1 August 2019

Abstract. The article is dedicated to the analysis of the International Commission of Continental Erosion (ICCE) of International Association of Hydrological Sciences (IAHS) activity since foundation in 1981. The empirical data is the collections of scientific articles, published in *IAHS Red Books* after symposiums and workshops, organized by the ICCE. 2460 authors published articles in *Red Books*, 81 % of them have publication in the one issue. The most active contributors (who have articles in 7 or more issues) compose a group of 39 scientists (1.6 % from the total number of authors). Researchers from 89 countries were published articles in the ICCE issues of *Red Books*. The activity of the ICCE can be described by a cyclic model of growth. It is shown that the ICCE scientific community is becoming progressively global.

1 Introduction

In a broad sense a scientific community is understood as a complex self-organizing system, which includes along with individual scientists, also various state institutions, public and commercial organizations, formal and informal groups, etc., which are interacting in the process of scientific activity both with one another and with external environment. The study of structure and functioning of scientific groups and communities is required to determine their role in the development of a certain scientific direction, to forecast the main trends of such development, and to increase the efficiency of management in scientific activities.

Traditionally the research subject in the area of scientific communities is a scientific laboratory, which is recognized to be the most important unit of collective scientific activity. There is abundant proof of studying a scientific laboratory as the main structure element of a scientific community (Pelz and Andrews, 1966; Knorr-Cetina, 1981, 1999; Woolgar and Latour, 1986). Typically, only sustainable and productive scientific groups can be taken as a research subject, since their history allows to trace the dynamics of their development during long-term time interval. The scientometric analysis of Makkaveev Scientific Laboratory of Soil Erosion and Channel Processes of Geographical Faculty of the Moscow State University can be shown as an example of such analysis (Bondarev and Boichenko, 2010, 2011). The described methodology of research can be used as a proven basis for further investigation of scientific groups and communities.

One of the least studied types of scientific communities are relatively stable international or interregional research groups that are created and operate in the process of interaction within regular conferences and meetings. As a rule, such conferences and meetings are organized by a small group of activists (both individuals and organizations), whose main purpose is to coordinate segmental research in a specific scientific area. The Interuniversity Scientific Coordination Council on the Problem of Erosion, Channel and Estuarial Processes at Lomonosov Moscow State University is an example of such scientific community (Bondarev and Rulyova, 2018).

The ICCE is another example of a sustainable and productive super-institutional scientific community. The goal of this study is to define basic scientometric characteristics of the ICCE functioning since the period of foundation in 1981.

2 Object of study

According to the information from the official website (International Commission on Continental Erosion, 2019) it is one of the Commissions of the International Association of Hydrological Sciences (IAHS). The ICCE fields of interests include the erosion, transport and deposition of sediment and associated nutrients and contaminants and the interaction of these processes with other components of the hydrological cycle and the environment. ICCE underlines the fundamental role of erosion and sediment delivery in many environmental issues, and supports the inter-disciplinary researches aimed to improve our understanding and management of such issues.

There are eight ICCE officers, who take the most active participation in supporting of the Commission activity. Since 1981 ICCE organize Symposia and workshops. The all oral and in some cases poster presentations were published in 37 IAHS "*Red Books*" and a few special issues of the *International Journals*, edited by members of ICCE group (ICCE Publications, 2019). These publications were taken as the main data set for the scientometric analysis.

3 Methods of study

Investigation of communicative interactions within scientific communities is the most important tool of scientometric analysis (Blinnikov, 1976; Woolgar and Latour, 1986; Ogurtsov, 2011). Latour and Woolgar (1986) were the first who proposed to identify of scientific groups by means of a special analysis of scientific communications. Professional scientific activity provides for extensive contacts, because process of scientific research presume that the scientist is in contact with many communication parties, each of which has its certain role in the process of accumulation and distribution of scientific knowledge. Therefore, it is impossible to assess the current condition and prospects for the development of a certain scientific group without the study of its communicative interactions.

Publication activity is commonly believed to be the most informative criteria for studying and evaluating a scientific group. Generally, the publication activity is understood as a statistical analysis of a designated publication array, selected to comply with the current research issues. The importance and applicability of this kind of research is based on the definition of scientific publication as the final result of both individual and collective scientific activities. The statistical analysis of publication array allows to make conclusions about: (a) the principals of labour specialization in science, based on the study of co-author relations; (b) the correlations in cross-disciplinary fields of study, based on thematic analysis; (c) the productivity of scientific group, based on the quantity and volume of publications, etc. (Dumenton, 1987). Thus, the publication array forms a specific data field of the particular scientific community, and may serve as the main subject for its scientometric analysis.

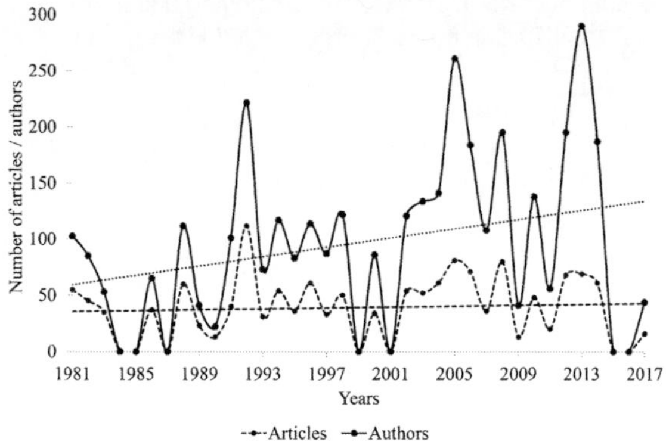

Figure 1. Dynamics of the number of participants in ICCE events: (a) – number of reports presented at the events; (b) – number of researchers who submitted as participants to the event.

4 Publication activity of the ICCE group members

ICCE arranges symposia and workshops on regular basis. The proceedings of the scientific meetings are published in IAHS "*Red Books*" and Special Issues of *International Journals* (for the purposes of the current study both types of publications will be hereafter named "the issues"). During the years of activity, IAHS Press has published 35 issues related to ICCE organized Symposia and workshops. To identify the common activity of ICCE members, the number of articles, article co-authors and countries, which they represented, have been counted for each issue. The individual issue for the time window 1981–2017 at an average included 39 articles written by 97 authors from 20 countries. The time fluctuations in the publication activity is presented on Fig. 1. It is possible to identify the two trends. Firstly, the general increase of the number of article co-authors This trend might result from the fact that nowadays scientific activity in the Earth Sciences become more collective due to increasing of number of methods and techniques, which are using for receiving final results Secondly, cyclic variability of number of articles in the issues. This trend can be explained the differences in the topics discussed during Symposia or workshop.

It is worth mentioning that during the whole period of ICCE activity, the most frequent number of co-authors is two. 88 % of the articles have from 1 to 4 authors. And while at the beginning of ICCE activity the tendency was more for one author per article, over recent years it is more typical, that number of co-authors are three per article (Fig. 2).

The activity of single authors and participating countries were also examined. There are several criterias in scientometrics, which allow to determine leaders in scientific communities. One of them is based on the well-known Pareto

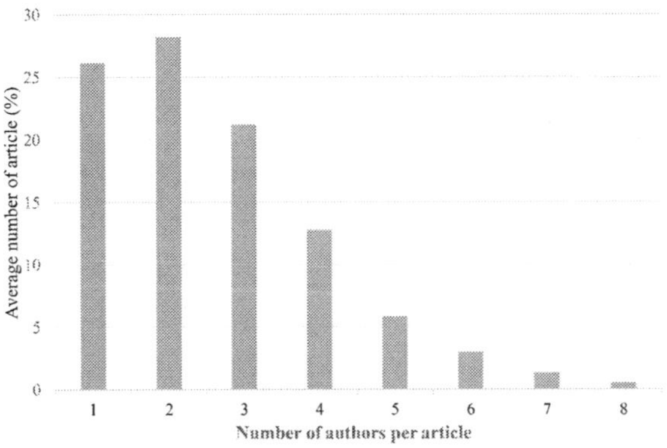

Figure 2. Average number of authors per article between 1981 and 2017.

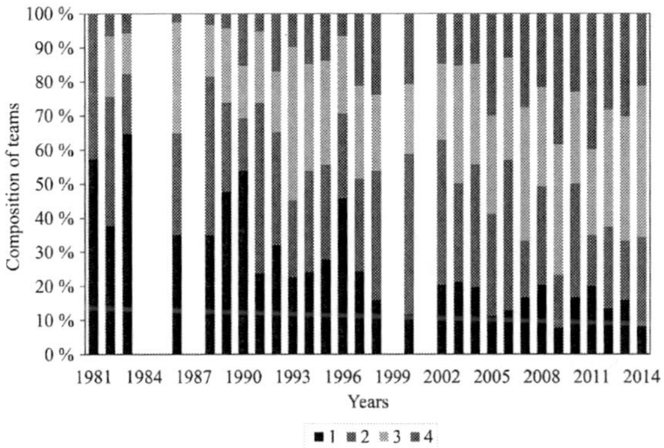

Figure 3. Time changes in the ratio of articles prepared with the participation of a different set of co-authors and published in ICCE edited issues. Legend: Articles prepared by: 1 – individual authors; 2 – co-authors from one organization; 3 – co-authors from one country (national team); 4 – international team.

principle, which states that, for many events, 80 % of the effects come from 20 % of the causes. The Pareto principle has a bearing on the most general trends and is widely used in economy and sociology. The statistical analysis of 37 issues allows us to reveal ICCE leaders among its regular members. The statistical analysis of 37 issues allows us to reveal ICCE leaders among its regular members. The total number of co-authors are 2460 scientists in total, 81 % from them (2000 authors) were co-author of publication in just one issue, whilst 19 % (460 co-authors) have had published articles in two or more issues. The most active scientists (who have participated in 7 or more issues) compose a group of 39 members or less than 1.6 % from the total number of authors.

As for the geography of participation, articles written by researchers from 89 countries were published in the ICCE issues during the examined time window 1981–2017. The absolute leaders in the number of co-authors are the UK, Germany, the USA, Canada and China (with Hong Kong). The scientists from the following countries are quite active in the participation and preparation of articles in the ICCE issues: Australia, France, India, Russia, Italy, Poland, Japan, Brazil, The Netherlands, Spain, Denmark, and Norway. There are 28 countries, which representatives took part in the preparation of articles for 8 or more issues of *ICCE Proceedings* (either *IAHS Red Books* or Special Issues of *International Journals*).

To analyze the structure of research teams the following data were examined: the number of articles with a single author, with co-authors from one organization, with co-authors from one country, with co-authors from different countries (Fig. 3). Unfortunately, data row ends in 2014, as the subsequent materials do not contain information concerning country of the article authors. It is possible to note the permanent reduction of the number of single authors. The number of co-authors from one organization remains quite stable, but the numbers of co-authors from different organizations and different countries are constantly increasing. The emerging and steady growth of the number of international teams witnesses that the process of globalization which influences all spheres of the society, has a distinctive impact on the structure of modern scientific groups. Inter alia the ICCE scientific community is becoming progressively global.

5 Conclusions

Scientometric analysis of the ICCE activity shows, that the ICCE scientific community has been actively developing to become at the moment "formed" and "stable" and to evidently have a strong potential for further growth. The statistics analysis of publication array allows to reveal a wide geography of its members, a significant number of participants, and an active core of high-profile scientists which allows this community to organize Symposia and workshops on the regular way. The analysis internal and external contacts, structure of research teams allows to conclude that the ICCE organizational framework and functioning are stable and adaptive enough to give this scientific community hopeful prospects for a sustainable future. The results of study can be useful both for the ICCE management and for its active participants, since they include original data revealing some special aspects of the ICCE structure and functioning as a super-institutional scientific community. Besides, the research data might be of interest to some specialists in the field of scientometrics and sociology of science.

Data availability. Data is available at https://doi.org/10.5281/zenodo.3262389 (Bondarev, 2019).

Competing interests. The author declares that there is no conflict of interest.

Special issue statement. This article is part of the special issue "Land use and climate change impacts on erosion and sediment transport". It is a result of the ICCE Symposium 2018 – Climate Change Impacts on Sediment Dynamics: Measurement, Modelling and Management, Moscow, Russia, 27–31 August 2018.

Acknowledgements. Processing of empirical data was carried out according to the plan of Research Laboratory of Soil Erosion and Channel Processes Lomonosov (Moscow State University), theoretical interpretation was carried out according to the research plan of Department of Sociology and Cultural Studies (Bauman Moscow State Technical University). The author is grateful to the following persons for their contribution: Valentin Golosov for his helpful advice and explanations of specific features of the ICCE structure and functioning, Olga Boichenko and Julia Yakubova for the assistance in text processing.

References

Bondarev, V.: Basic scientometric characteristics of ICCE, Zenodo, https://doi.org/10.5281/zenodo.3262389, 2019.

Bondarev, V. and Boichenko, O.: Structure and Dynamics of scientific Lab staff, Sotsiologicheskie Issledovanya [Sociological Studies], 11, 52–63, 2010 (in Russian).

Bondarev, V. and Boichenko, O.: The structure and functioning of the research team (communicative aspect), MSU Vestnik, 18, Sociology and Political Science, 1, 80–97, 2011 (in Russian).

Bondarev, V. and Rulyova, S.: Structure and Dynamics of Meta-Organizational Scientific Communities, Sotsiologicheskie Issledovanya [Sociological Studies], 5, 32–144, https://doi.org/10.7868/S0132162518050124, 2018 (in Russian).

Blinnikov, L. V. (Ed.): Communication in modern science, Moscow, Nauka, 1976 (in Russian).

Dumenton, G. G.: The Network of scientific communication and organization of fundamental research, Moscow, Nauka, 366 pp., 1987 (in Russian).

ICCE Publications: International Commission on Continental Erosion, available at: http://www.icce.altervista.org/icce-publications.html/, last access: 15 May 2019.

International Commission on Continental Erosion: available at: http://www.icce.altervista.org/, last access: 15 May 2019.

Knorr-Cetina, K.: The manufacture of knowledge: an essay on the constructivist and contextual nature of science, Oxford, New York, Pergamon Press, 204 pp., 1981.

Knorr-Cetina, K.: Epistemic cultures: how the sciences make knowledge, Cambridge, Massachusetts, Harvard University Press, 352 pp., 1999.

Ogurtsov, A. P.: Scientific discourse: power and communication (complementarity of the two traditions), Philosophy of science: the twentieth century, St. Petersburg, 2, 12–59, 2011 (in Russian).

Pelz, D. C. and Andrews F. M.: Scientists in Organizations: Productive Climates for Research and Development, Oxford, England, John Wiley, 393 pp., 1966.

Woolgar, S. and Latour, B.: Laboratory life: the construction of scientific facts, Princeton, New Jersey, Princeton University Press, 296 pp., 1986.

Proc. IAHS, 381, 7–11, 2019
https://doi.org/10.5194/piahs-381-7-2019

Integrated approach to study river fluxes, water and sediment sources apportionment in sparsely monitored catchment

Sergey Chalov[1], Pavel Terskii[1], Thomas Pluntke[2], Ludmila Efimova[1], Vasiliy Efimov[1], Vladimir Belyaev[1], Anna Terskaia[1], Michal Habel[3], Daniel Karthe[4], and Christian Bernhofer[2]

[1]Department of Geography, M.V. Lomonosov Moscow State University, Moscow 119991, Russia
[2]Technische Universität Dresden, Chair of Meteorology, Dresden, Germany
[3]Institute of Geography, Kazimierz Wielki University, Bydgoszcz 85-064, Poland
[4]Engineering Faculty, German-Mongolian Institute for Resources and Technology, Nalaikh, Mongolia

Correspondence: Sergey Chalov (hydroserg@mail.ru)

Published: 1 August 2019

Abstract. The paper presents an integrated approach for the apportionment of water and sediment sources and pathways in ungauged territories. The approach comprises various elements such as a large-scale semi-distributed hydrological and erosion model (using the SWAT tool), a statistical assessment of regular discharge data, the discrimination of runoff sources using end-member mixing analysis (EMMA), the apportionment of sediment sources using fingerprinting technique and sediment concentration and water discharge hysteresis rating curves analyses. The approach has been applied to Western Dvina River Basin (WDRB). The results discussed in the paper indicate main features of the water and sediment regime. The components of the presented framework provide a guide for detailed water and sediment studies making it an inexpensive way to develop a roadmap to understand important hydrological processes at a watershed scale.

1 Introduction

Riverine fluxes, which represent the amount of chemicals transported along the river, are an important subject in rivers studies. Many rivers face problems related to the uneven spatial distribution of monitoring stations. In transboundary rivers, this is mostly due to discrepancies in monitoring programs of the riparian countries. The latter is being the main cause of significant uncertainties (Alexeevskii et al., 2013; Heininen, 2018) and gaps in knowledge on water and sediment quality of Western Dvina River Basin (WDRB), especially its upper part, which is located in Russia. The Western Dvina has its source in the Valdai Hills, Russia, and then flows just above 1000 km through Russia, Belarus, and Latvia, where it finally drains into the Gulf of Riga of Baltic Sea. Even though the Western Dvina does not flow through either country, parts of Lithuania and Estonia drain into the WDRB. The gauging network is sparse and data has a lot of gaps, especially related to water and sediment quality (Chalov et al., 2017b). It is characterized by six stations representing a drainage area of 17 000 km^2 with water discharges monitoring and one single sediment monitoring station at the outlet of the catchment. Testing independent approaches becomes an essential part of sediment studies in such kind of areas (e.g. Chalov et al., 2017a). According to WMO (2008), the recommended minimum station density for gauging stations is between 1 station per 1000 km^2 in mountaineous subcatchments, 1 station per 1875 km^2 in hilly areas and interior plains, and 1 station per 2750 km^2 in coastal areas.

2 Description of integrated approach

Multidisciplinary ecohydrological approach at Russian part of WDRB has been established to make a first study of water and sediment transport processes within a territory (Table 1). It is based on integrating independent field and numerical tools (modules) to develop a comprehensive conceptual

Table 1. Integrated approach to study water and sediment transport.

Research scope	Module		
	Type*	Water	Sediments and tracers
A. Sources	I	1. Field reconnaissance and land cover classification 2. Water and sediment sources sampling 3. Experimental study on stream flow-river bed mass exchange	
	II	1. Discriminating runoff sources using end-member mixing analysis (EMMA)	2. Discriminating sediment sources using fingerprinting technique 3. Sediment concentration – water discharge hysteresis rating curves analyses
B. Hydro-logical pathways	I	1. Water flux assessment from regular discharge datasets	2. Spatio-temporal assessment of water quality in the WDRB catchment 3. Sediment and tracers flux analyses from regular measurements datasets
	II	1. Long term data for flow discharge (Q) from available gauges (daily and monthly) 2. Large scale semi-distributed hydrological and erosion model (SWAT): 2a. meteorological ERA-Interim reanalysis 2b. catchment delineation 2c. land cover and soil classification 3. Small scale detailed semi-distributed hydrological model (SWAT)	3. Long term data for sediment discharge (S) from downstream station (Velezh) (event based) using statistical sediment load model (LOADEST) 4. Semi-distributed erosional model using hydrological SWAT model

* Field-based (I) or numerical and remote sensing (II) modules.

model of the sources and transport of water, sediment and tracers. Two principal research scales are concerned: sources (A) and hydrological pathways (B) (Table 1). The development of the process-based water, sediment and tracers dynamic integrated approach for pilot study in Western Dvina Basin involved several field-based (I) or numerical and remote sensing (II) modules. The field based studies consisted of field reconnaissance, water and sediment sources sampling, experimental study on stream flow-river bed mass exchange (all related to AI, Table 1) and water flux assessment from regular discharge datasets, spatio-temporal assessment of water quality in the WD catchment, sediment and tracers (total dissolved load) flux assessment from regular discharge datasets obtained (all related to BI, Table 1).

The numerical modules (II) included field-based and remote sensing monitoring program to determine the principal factors of water, sediment and tracers dynamics, application of distributed hydrological model SWAT v.2012 was chosen (Arnold et al., 2012) as a tool to simulate topography, soil and LULC data impact on water balance. The advantages of the SWAT model are the availability of reliable and helpful documentation, absence of limitations on catchment area, compatibility with GIS software (ArcGIS, QGIS, MapWindow), the open-source module of auto calibration SWAT-CUP (Abbaspour et al., 2007). The modeling process of different types of fluxes (water, dissolved and suspended matter) is based on calculations of the river discharge and its parametrization. Based on the catchment inventory, development of soil erosion, sediment transport and deposition modules for channel and hillslope areas separately can be done using statistical model LOAD ESTimator (LOADEST), a FORTRAN program for estimating constituent loads in streams and rivers (Runkel et al., 2004) and integration of these modules with SWAT consistently (1). The models calibration additionally uses results of field-based catchment land cover and soils description and classification. Each of the modules was numbered in 3-digit format (A, B – research scope (sources or pathways); I, II – type (I – field-based; II – numerical and remote sensing); 1, 2, 3, 4, 5 – exact tool).

Below we discuss the main results of three modules of the integrated approach – A-II-1, A-II-2 and B-II-2,3. In particular, A-II-1 is aimed to study runoff generation based on hydrochemical assessment of interaction of waters from various water sources of a river in the process of mixing within the river basin based on the samples collected as a part of A-I-2 module (Table 1). A-II-2 module is focused on the identification of the representative sources of sediment and is based on selected tracers (Hooper et al., 1990). B-II-2,3 is based on hydrological model software application of SWAT) (Krengel et al., 2018). Sediment yield in SWAT is estimated with the modified soil loss equation (MUSLE) developed by Williams and Berndt (1972), which is a modification of

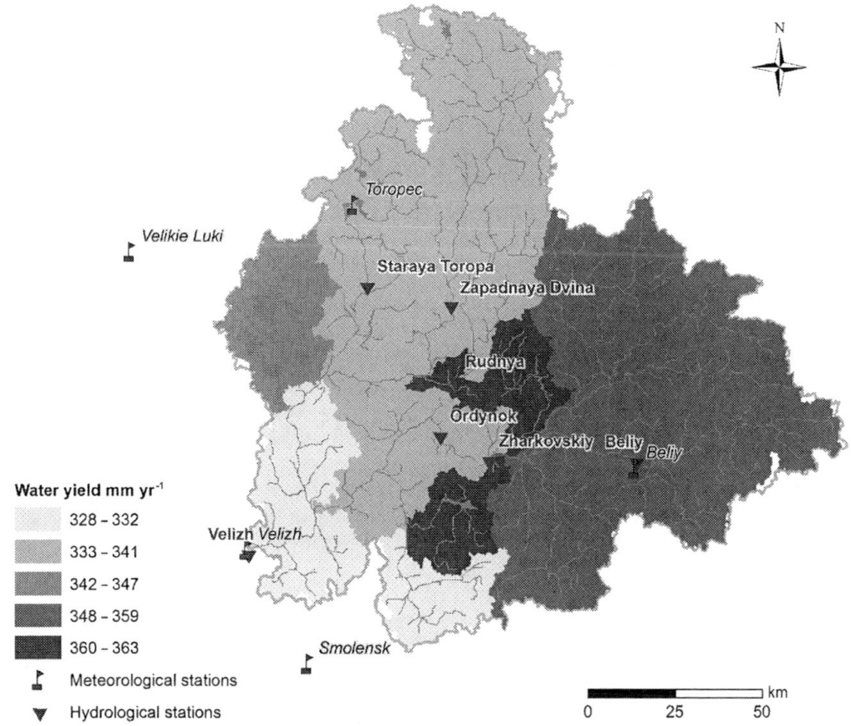

Figure 1. Annual water yield distribution for Russian part of WDRB (SWAT based – mean for 1992–2016).

the standard Universal Soil Loss Equation of Wischmeier and Smith (1960). The results of the field-based catchment land cover and soils description and classification are used for model calibration (modules A-1). Additionally, the land cover classification based on geobotanical descriptions consists of textual vegetation type description. It is used for an automated interpretation of remote sensing images. All images were classified separately using maximum likelihood supervised classification and then merged into mosaic. Initially, 45 classes were produced that were further grouped manually into 12 classes according to NLCD 92 classification (Vogelmann et al., 2001), which can directly be used in SWAT model.

3 Results and discussion

The integrated approach (Table 1) was tested as a tool to characterize water and sediment processes within the sparsely monitored WDRB basin. A comparison of meteorological observation data and modelled re-analysis data was undertaken. We recommend to use the interpolated observation data, except the downward solar radiation, because no observations exist. It was investigated, whether a monthly or a daily time step is more appropriate to achieve satisfactory results for the hydrological modelling. Using a detailed digital elevation model DEM (12.5 m) and land use/land cover (30 m) significantly improve results for a daily time step, but almost does not have any effect for monthly. A sensitivity analysis identified the most important calibration parameters: some "snow" and "groundwater" parameters, and the distributed "rainfall-runoff curve number" parameter. The calibration of "snow" parameters should be done separately, because of the different genesis of snowmelt processes against infiltration and evapotranspiration. Five "snow" parameters were set to naturally adequate ranges for the auto calibration procedure SWAT CUP (Abbaspour et al., 2007). Daily runoff is simulated good (Moriasi et al., 2007) by Nash-Sutcliff efficiency, but annual mean is overestimated (PBIAS reaches 16 %). Evaporation is simulated well, but snow water equivalent is slightly overestimated (in comparison to observed). We found, that the soil database should be more detailed for daily time step calculations. For the period of model setup 1992–2016, water balance components were calculated: snow melt, actual evapotranspiration, soil water content, percolation, surface runoff, groundwater and lateral flow, total water yield (Fig. 1). Spatial distribution of all simulated water balance components represents the structure of flow generation drivers inside the catchment. It strongly depends on land cover and soil types. The highest evapotranspiration is linked to sub basins with lakes and wetlands predomination in the western part of the catchment. Higher snowmelt is linked to higher precipitation in the eastern part. Higher soil water and lower lateral and groundwater flow are linked to gleyic soils and histosols which are dominant under wetlands. The objectives of the EMMA analysis done in the Velesa River basin (module WP A-II-1) were to identify and interpret the potential sources of water in WDRB. The preliminary results indicate a significant variability of underground water sources. The chemical composition of these sources is determined by soil-formation processes and the underlying rocks. The identification and accounting of specific water sources in the EMMA-analysis will make it possible to give an objective assessment of the temporal dynamics in dominant runoff sources and flow paths.

The preliminary results of suspended sediment sources appointment using fingerprinting techniques (working package A II-2) within the Velesa River basin provides valuable information on spatial and temporal variability of the contribution of different sediment sources to the total suspended load. Accordingly, a sediment source fingerprinting technique was used to assess the spatial sources of fine-grained ($< 10\,\mu$m) sediments. Total content of selected elements and 137Cs radioactive isotope was used for discriminating sources to compare it with time-integrated suspended sediment samples (TISSS). The results suggest that the main contributors of fine-grained sediment are the intensively eroded river banks. The results of the sediment source apportionment are in accordance with earlier findings for the rivers of similar size from the forest zone (e.g. Theuring et al., 2013, 2015). According to the obtained results, river bank erosion generates over 70 % of the total fine-grained sediment load. The use of a combination of independent statistical tests helps to take account of potential uncertainties in source estimates, associated with the use of a single composite signature. We found that the application of mobile forms of elements for fingerprinting purposes is questionable, because these may not meet the criteria of conservativeness. Active gullies and river channel erosion sections like undercut banks should be determined in detail based on topographic maps and remote sensing data, which are validated with field data.

This study examines the development of a conceptual framework of water and sediment processes in the catchment scale. Spatially distributing the components of a approach provides a guide for developing more detailed water and sediment budgets. It is inexpensive way to develop a roadmap for understanding the linkages among the various transport processes at a watershed scale. The approach is particularly aimed to provide a full quantitative description of river fluxes and sediment and water quality in poorly gauged catchments. This is very important in areas, where no or only limited sediment-related studies have been done in the past. We found that almost no studies have been devoted to ecohydrology, hydromorphology and sediment transport in transboundary catchments of Eastern Europe (Chalov et al., 2017b; Krengel et al., 2018). The proposed approach might be a useful tool to fill these knowledge gaps.

Data availability. Hydrological data from Russian State Hydrometeorological Agency was used in the study which is the property of this organization. Additionally, future publications of the authors are intended to publish field datasets mentioned in this paper.

Author contributions. SC designed a study concept, CB is the project leader and help to develop the concept of the approach. PT and AT are responsible for SWAt model and satellite data procesing, TP was part of the SWAT modelling group and prepared the climate data. DK made a general text writing and editing, LE and VE have done the water quality study, VB run the fingerprinting model, MH provided assistance for data analyses.

Competing interests. The authors declare that they have no conflict of interest.

Special issue statement. This article is part of the special issue "Land use and climate change impacts on erosion and sediment transport". It is a result of the ICCE Symposium 2018 – Climate Change Impacts on Sediment Dynamics: Measurement, Modelling and Management, Moscow, Russia, 27–31 August 2018.

Acknowledgements. The authors thank the Volkswagen Foundation for the funding of the project "Management of Transboundary Rivers between Ukraine, Russia and the EU – Identification of Science-Based Goals and Fostering Trilateral Dialogue and Cooperation (ManTra-Rivers)".

Financial support. This research has been supported by the Volkswagen Foundation (grant no. 90 426). Apportionment methodology is developed as a part of Russian Science Foundation project no. 19-77-30004.

References

Abbaspour, K. C., Vejdani, M., and Haghighat, S.: SWAT-CUP calibration and uncertainty programs for SWAT, Proc. Intl. Congress on Modelling and Simulation (MODSIM'07), 73, 1603–1609, 2007.

Alekseevskii, N., Zavadskii, A., Krivushin, M., and Chalov, S.: Hydrological Monitoring at International Rivers and Basins, Water Resour., 42, 747–757, 2015

Arnold, J. G., Moriasi, D. N., Gassman, P. W., Abbaspour, K. C., White, M. J., Srinivasan, C., Santhi, R., Harmel, D., Van Griensven, A., Van Liew, M. W., Kannan, N., and Jha, M. K.: SWAT: Model use, calibration, and validation, T. ASABE, 55, 1491–1508, https://doi.org/10.13031/2013.42256, 2012.

Chalov, S., Golosov, V., Tsyplenkov, A., Theuring, P., Zakerinejad, R., Märker, M., and Samokhin, M. A.: toolbox for sediment budget research in small catchments, Geogr. Environ. Sustain., 10, 43–68, https://doi.org/10.24057/2071-9388-2017-10-4-43-68, 2017a.

Chalov, S., Pluntke, T., Nabyvanets, Y., Kruhlov, I., Helm, B., Terskii, P., Krengel, F., Osadcha, N., Habel, M., Osadchyi, V., Reeh, T., Karthe, D., and Bernhofer, C.: Report on WP A – Assessment of the status quo in the three investigated basins, Interim Report of the Volkswagenstiftung project "Management of Transboundary Rivers between Ukraine, Russia and the EU ...", 10 pp., 2017b.

Heininen, L.: Arctic geopolitics from classical to critical approach – importance of immaterial factors, Geogr. Environ. Sustain., 11, 171–186, 2018.

Hooper, R. P., Christophersen, N., and Peters, N. E.: Modelling streamwater chemistry as a mixture of soilwater end-members – an application to the Panola Mountain catchment, Georgia, USA J. Hydrol., 116, 321–343 https://doi.org/10.1016/0022-1694(90)90131-G, 1990.

Krengel, F., Bernhofer, C., Chalov, S., Efimov, V., Efimova, L., Gorbachova, L, Habel, M., Helm, H., Kruhlov, I., Nabyvanets, Y., Osadcha, N., Osadchyi, V., Pluntke, T., Reeh, T., Terskii, P., and Karthe, D.: Challenges for transboundary river management in Eastern Europe-Three case studies, Die Erde, 149, 157–172 https://doi.org/10.12854/erde-2018-389, 2018.

Moriasi, D. N., Arnold, J. G., Van Liew, M. W., Bingner, R. L., Harmel, R. D., and Veith, T. L.: Model evaluation guidelines for systematic quantification of accuracy in watershed simulations, Am. Soc. Agr. Biol. Eng., 50, 885–900, 2007.

Runkel, R. L., Crawford, C. G., and Cohn, T. A.: Load Estimator (LOADEST): A FORTRAN Program for Estimating Constituent Loads in Streams and Rivers: US Geological Survey Techniques and Methods Book 4, chap. A5, 69 pp., 2004.

Theuring, P., Rode, M., Behrens, S., Kirchner, G., and Jha, A.: Identification of fluvial sediment sources in the Kharaa River catchment, Northern Mongolia, Hydrol. Process., 27, 845–856, https://doi.org/10.1002/hyp.9684, 2013.

Theuring, P., Collins, A. L., and Rode, M.: Source identification of fine-grained suspended sediment in the Kharaa River basin, northern Mongolia, Sci. Total Environ., 526, 77–87, https://doi.org/10.1016/j.scitotenv.2015.03.134, 2015.

Vogelmann, J. E., Howard, S. M., Yang, L., Larson, C. R., Wylie, B. K., and Van Driel, N.: Completion of the 1990s National Land Cover Data Set for the conterminous United States from Landsat Thematic Mapper data and ancillary data sources, Photogramm. Eng. Rem. S., 67, 650–662, 2001.

Williams, J. R. and Berndt, H. D.: Sediment yield computed with universal equation, J. Hydraul. Div., 98, 2087–2098, 1972.

Wischmeier, W. H. and Smith, D. D.: A universal soil-loss equation to guide conservation farm planning, Trans 7th int Congr. Soil Sci., 1, 418–425, 1960.

WMO: World Meteorological Organization Guide to Hydrological Practices: Volume I Hydrology – From Measurement to Hydrological Information, 6th Edn., Geneva, Switzerland, World Meteorological Organization Publication No. 168, 296 pp., 2008.

Proc. IAHS, 381, 13–20, 2019
https://doi.org/10.5194/piahs-381-13-2019

Lake sedimentation as an agent of postglacial transformation of interfluves and fluvial landscapes of the Borisoglebsk Upland, Central European Russia

Ekaterina V. Garankina[1], **Vladimir R. Belyaev**[2], **Ilya G. Shorkunov**[3], **Yuliya V. Shishkina**[1], **Pavel V. Andreev**[4], and **Elena D. Sheremetskaya**[1]

[1]Department of Geomorphology and Paleogeography, Faculty of Geography, Lomonosov Moscow State University, Moscow, 119991, Russia
[2]Laboratory for Soil Erosion and Fluvial Processes, Faculty of Geography, Lomonosov Moscow State University, Moscow, 119991, Russia
[3]Department of Soil Geography and Evolution, Institute of Geography, Russian Academy of Sciences, Moscow, 119017, Russia
[4]Institute of Earth Sciences of Saint-Petersburg State University, St Petersburg, 199034, Russia

Correspondence: Ekaterina V. Garankina (evgarankina@gmail.com)

Published: 1 August 2019

Abstract. Borisoglebsk Upland is considered an example of a secondary upland plain in the marginal zone of the last Middle Pleistocene glaciation. Moraine hills and kames were reworked by glaciofluvial processes and incised by small fluvial forms later. Its postglacial surface drift cover is regularly defined as an undivided complex of mantle loams of dominantly subaerial origin with characteristic cryogenic features and remnants of paleosols. However, some previous studies suggest that lake sedimentation played an important role in the postglacial history of the Borisoglebsk Upland. This paper presents results of a detailed investigation of postglacial sedimentary cover of the eastern part of the Borisoglebsk Upland aimed to reconstruct the co-evolution of surface deposits, soil cover and geomorphic landscapes since degradation of the last Middle Pleistocene glaciation about 150 ka (MIS-6). The study is essentially based on a comprehensive lithological, pedological and geocryological description of postglacial deposits in cores (hand or machine-driven) and open sections, systematic sampling for grain size analysis and selective sampling for ^{14}C absolute dating and monoliths structural examination. The results indicate that most of the surface drifts in this feature consists of stratified lacustrine deposits. Their Late Pleistocene age is stratigraphically confined by the underlying paleosols and incorporated peats of the Mikulino interglacial age (MIS-5) and several organic-rich layers within the lake sequence ^{14}C dated to the Middle Valdai interstadial (MIS-3). Overlying mantle loams and colluvial deposits with cryogenic features and low organic matter content those facially substitute lacustrine sediments were attributed to the Late Valdai stadial (MIS-2). After the Mid-Holocene stabilization, relatively thin colluvial cover identified by the increased amount of organic matter also deposited. We conclude that lacustrine sedimentation is the primary Late Pleistocene agent that transformed the initial glacial topography and most characteristic type of lithodynamics of the eastern Borisoglebsk Upland.

1 Introduction

The eastern part of the Borisoglebsk Upland is located near the Nero Lake in Central European Russia and has long been an attraction for paleogeographers and archeologists who conduct research to reconstruct the natural environment and human settling histories. However, the detailed evolution of this landscape during the postglacial period remains unclear. In particular, little is known about the formation of interfluve areas and smaller elements of the fluvial network. Classically this region is regarded as a secondary upland plain in the marginal zone of the last Middle Pleistocene glaciation (Moscow Age, MIS-6) (Semenov et al., 1972; Astakhov et al., 2016). Topography of this feature include moraine hills and kames reworked by glaciofluvial activity and partly dissected by small valleys and gullies. Surface drift cover overlying glacial tills is often defined as an undivided complex of mantle loams. A complete genetic classification of these silty and sandy loams with characteristic cryogenic features as well as remnants of paleosols is currently lacking. Typically not exceeding a depth of 2–3 m, they are reported across several parts of European Russia and Europe and are correlated with loesses of the southerly regions (Velichko et al., 2006). Many researchers propose this feature is predominantly of subaerial origin formed under periglacial conditions that included a combination of aeolian, cryogenic and shallow pool sedimentation processes (Makeev et al., 2017). However, geological map of Quaternary deposits (O-37-XXVIII, scale 1 : 200 000) indicates that this upland consists of up to 8 m of mantle loams (Semenov et al., 1972). Maximum relative elevation between the highest interfluve and the Nero Lake level is about 111 m (Fig. 1b). However, local topographic ranges are mostly within 5–20 m, which suggests the thickness of surface loams to be close to the relative elevation of local landforms. Accordingly, detailed information on its origin is an important key to further elucidate evolution of the geomorphic landscape.

Individual studies suggest that lacustrine sedimentation played an important role in the postglacial history of the Borisoglebsk Upland (Kvasov, 1975; Rusakov and Sedov, 2012). Our recent detailed analysis of these postglacial deposit cover shows that a significant part is composed of lake sediments that are unrelated to the main Nero Lake terraces (93–120 m a.s.l.). Moreover, these lacustrine deposits not only cover the most recent glacial tills but they also play an important part in the modern topography of both lower interfluvial surfaces and upper parts of gullies and small valleys in elevation range of 150–180 m a.s.l.

The present paper provides new insights regarding the lithology, paleopedology and relic cryogenic deformation of the Late Pleistocene surface drifts located on the eastern macroslope of the Borisoglebsk Upland. Accordingly, the results of this investigation are employed to reconstruct the co-evolution of surface deposits and geomorphic landscapes of the study area since degradation of the last Middle Pleistocene glaciation about 150 ka.

Table 1. Sampling protocol of the 2016–2019 field campaigns.

Type of analysis	Number of samples collected	Number of samples analyzed
Grain size	434	150
LOI	338	83
^{14}C dating (AMS/scintillation)	25/22	3/0
Core monoliths	53	25

2 Materials and methods and study site description

The study is based on a comprehensive lithological, pedological and geocryological description of postglacial deposits in cores (hand or machine-driven) and open sections followed by characterizing the grain size and loss on ignition of bulk samples and in monoliths, inspecting of microstructures and selective ^{14}C dating. Core positions were obtained by post-processing differential GNSS survey with Leica GX 1200 base-rover complex with typical horizontal error $<\pm 2$ cm and vertical $<\pm 10$ cm. Topography was analyzed by means of both visual field descriptions, GNSS profiling, conventional geomorphic mapping and the GIS-based morphometric analysis of open-source satellite imagery (DigitalGlobe/GeoEye (Google Earth), IRS and LANDSAT (Yandex Maps), global satellite radar DEMs (SRTM and ALOS 3D), land planning topographic maps (1 : 10 000) and aerial photography obtained by UAVs DJI Phantom III and IV).

Grain size analysis involved standard sample preparation (Konert and Vandenberghe, 1997), dry sieving of coarser fractions (> 100 µm) by the Fritsch Analysette 3 PRO mechanical sieve shaker and laser diffractometry of the finer component (< 100 µm) by the Fritsch Analysette Nanotek 22 at the Laboratory of Experimental and Dynamic Geomorphology and Paleogeography, MSU, Russia (Table 1). Organic and chemical carbon content was measured by ignition at 500 and 900 °C (ASTM D 2974-87, 1993; Heiri et al., 2001). Structural organization and textural composition of sedimentary and soil (paleosol) bodies were examined using macro-photography in reflected light (macromorphology for more than 230 m and mesomorphology for about 100 m of cores and sections). Organic-rich layers were collected for ^{14}C dating using the AMS technique at the Laboratory of Radiocarbon Dating and Electronic microscopy, Institute of Geography RAS, Russia, and Center for Applied Isotope Studies, University of Georgia, USA, and scintillation technique at the Radiocarbon Dating Laboratory, Institute of Environmental Geochemistry, National Academy of Sciences, Ukraine (Table 2).

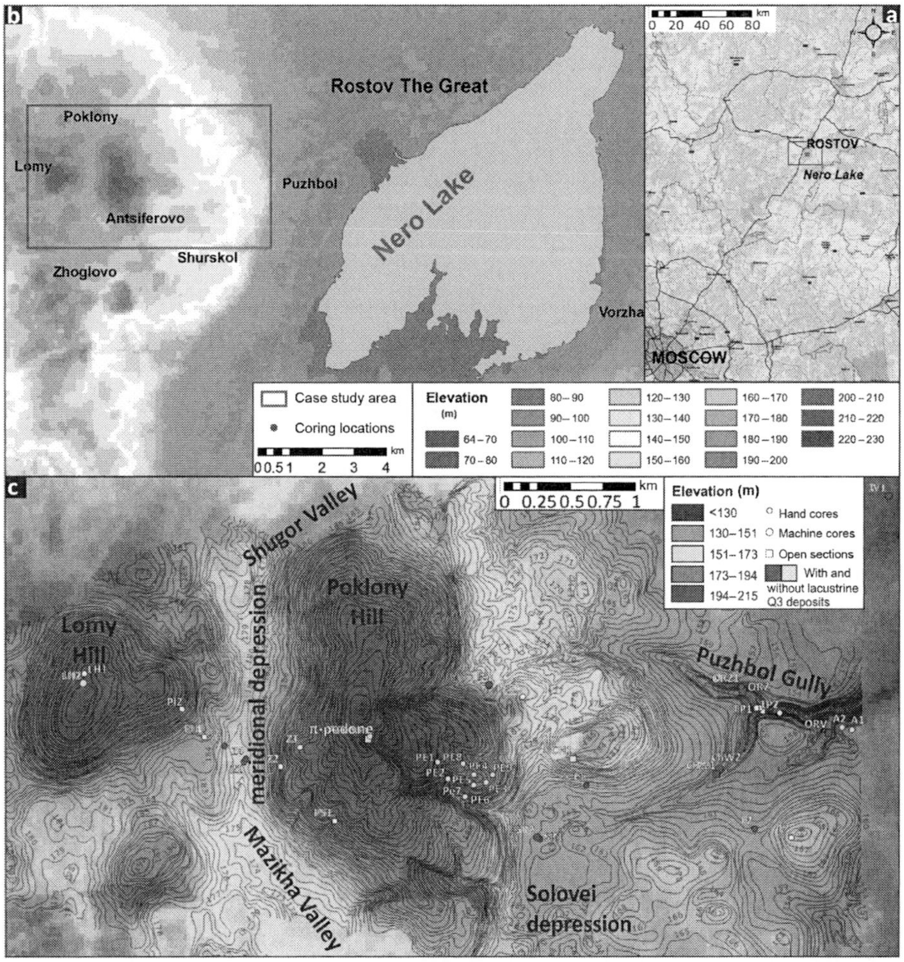

Figure 1. Position of the case study area: **(a)** at the Central Russian Plain (source: ©Google global topography, infrastructure and settlements), and **(b)** at the eastern Borisoglebsk Upland (source: ©ALOS 3D 30 m); **(c)** its detailed topography based on 1 : 10 000 land planning maps (contour interval 1 m).

Borisoglebsk Upland is located in the center of the Russian Plain about 200 km to the northeast from Moscow City, in the southern part of the Yaroslavl Region (Fig. 1a). Historically this landscape was considered as a secondary Moscow glacial plain (MIS-6) moderately reworked by superimposed, primarily glaciofluvial and fluvial, processes (Novskij, 1975; Semenov et al., 1972; Sudakova et al., 2013). In the study area (Fig. 1b), most elevated (up to 214 m a.s.l.) interfluves and regional base level (Nero Lake, 93 m a.s.l.) are closely placed determining a specific rolling-stepped structure of local topography. It is incised by small valleys and associated gully systems.

The pedolithostratigraphy of the study area located north of Shurskol and Puzhbol settlements was investigated from interfluves down to the Nero Lake terraces (the lowest studied section at 105 m a.s.l.). Below 180 m a.s.l., the gentle slope is composed of thick (up to 10 m) complex deposits overlying the irregular surface of the Middle Pleistocene glacial till. Its structure was examined in 35 hand (up to 8 m deep, core diameter 3 cm) and 9 machine (up to 20 m deep, core diameter 8 and 10 cm) cores (total depth > 250 m) as well as 9 pits, quarries or natural exposures (Fig. 1c). Several sites previously studied by Semenov et al. (1972), Lavrushin, Chistyakova (2001), and Rusakov et al. (2015) were further investigated by our research group (Shishkina et al., 2019). Sampling strategy was modified to separate initial patterns of lithological composition and superimposed cryo- and pedofeatures.

Table 2. AMS ^{14}C dating results from the meridional depression key site and corresponding calibrated ages.

No.	Depth/altitude, m	Core	Layer	Material	Lab number	^{14}C, yr BP ($\pm 1\sigma$)	cal yr BP	
							2σ	Median
1	0.54–0.56/176.9	Z4	upper layer of cover beds	TOC	IGAN_{AMS} 6289[a]	3630 ± 20	3877–4062	3941
2	4.13–4.22/173.8	Z1	base of alluvium	Plant rs.	IGAN_{AMS} 6290[a]	$40\,680 \pm 310$	43 535–44 864	44 228
3	5.52–5.56/173.8	Z5	lower part of lacustrine unit	TOC	IGAN_{AMS} 6288[a]	$32\,450 \pm 155$	35 958–36 750	36 335
4	0.96–1.02/160.8	SK2	gyttja layer in slope deposits	TOC	Ki-19 676[b]	5190 ± 60	6179–5754	5955
5	5.50–5.55/156.3	SK2	peat lens in the middle part of lacustrine unit	TOC	Ki-19 671[b]	$31\,440 \pm 300$	36 007–34 746	35 336

[a] Results were calibrated using ^{14}C date calibrating programs CALIB REV7.1.0 and [b] OxCal v4.3.2.

3 Results and discussion

The interfluve topography of the eastern Borisoglebsk Upland is comprised of two types of hill structures with strikingly different morphology. Interfluves at an elevation of 180–215 m a.s.l. are represented by two large (25–50 m relative elevation, basal diameter up to 1.5–3.0 km) flat-topped hills with comparatively steep and in some places gradually stepped, slopes. They are separated by the glaciofluvial depression stretched in almost longitudinal direction and connecting headwaters of the two modern small valleys (Fig. 1c). 17 cores and an open section at the highest hilltop show widespread distribution of thick (> 10 m) contrastingly stratified clayey loams and sands covered by less clearly fine-laminated silty loams 1–2.5 m deep. That allowed us to propose a glaciolacustrine hypothesis of origin of such larger hills. They were formed by material from lakes initially located on top or inside of the melting Moscow glacier superimposed on the glacial till surface – kame-like mechanism of formation. Lower interfluves surrounding the base of these larger hills exhibit some evidence of the dead ice topography. Groups of shallow rounded depressions and low hillocks with relative elevation not exceeding 4–5 m (so-called kettle holes topography) are isometric or slightly elongated (Garankina et al, 2018).

The morphology and geological structure of the lower interfluves (180–130 m a.s.l.) between the basis of the highest hills and the upper Nero Lake terraces were examined in 24 cores and 4 exposures on gully sides. At all locations, Moscow glacial tills are covered by a pronounced fine-grained layer (silty to clayey loams with fine sandy enrichments) with clearly visible and texturally contrast multiscale lamination (up to five levels of laminae thickness, Fig. 2). Mottled color alternations (rusty yellow to pinkish or blueish grey to brownish dark grey) are indicators of FeO_3 and FeO_2 forms and organic matter that are typical of deposits created in standing water. Organic-rich lenses (peats, plant detritus or paleosols) and visible signs of relic pedogenesis (desiccation pores and planes) clearly identify the interruptions in lacustrine sedimentation. Simultaneously levels with well-developed cryogenic deformations (Belyaev et al., 2019) or traces of slope erosion material (subangular clayey clasts, Fig. 2g) point towards the periods of shallow lake levels.

Upwards, sediments tend to lighten in color with amplifying yellowish hue and lacking of any organic matter. Lamination is also less contrast and clear. With more or less distinct contact, they are substituted by deposits variegated in color. Those patchy or lenticular rather than regularly laminated textures with highly contrasting lithology interspersed with sands and gravels suggest their primary subaerial origin due to intensified solifluction and slope erosion and distinct trend of lakes to cease existing. Such phenomenon was described for the Central European region as periglacial cover beds (Kleber and Terhorst, 2013; Reinhardt-Imjela et al., 2018). We suggest a potential to correlate the intermediate and upper layers of cover beds with the upper lacustrine facies with strong colluvial income and overlying slope deposits of the study area.

The highest upper boundary of Late Pleistocene lacustrine deposits (177.6 m a.s.l.) was in Z5 core and its lowest base (133.4 m a.s.l.) – in Iv1 core (Fig. 1b). Hence, the elevation range where Late Pleistocene lacustrine deposits were observed (beyond the limits of Nero Lake terraces) is almost 45 m. In general, they are found in almost all exposures below 180 m a.s.l. except for the hilltops those are composed of 2–4 m of very fine subaerial deposits – silty loams to clays (core F3, A1).

SK2 core (161.8 m a.s.l.) with the thickest Late Pleistocene lacustrine deposits is located at the bottom of the Solovev depression (Fig. 1c). More than 6 m of fine-grained material (dominantly sandy to clayey loam) with some input from adjacent slope gullies (sand to small gravel particles) alternates with and is underlain by peat layers. Plant macrofossils emerge in almost the entire volume of the core ampli-

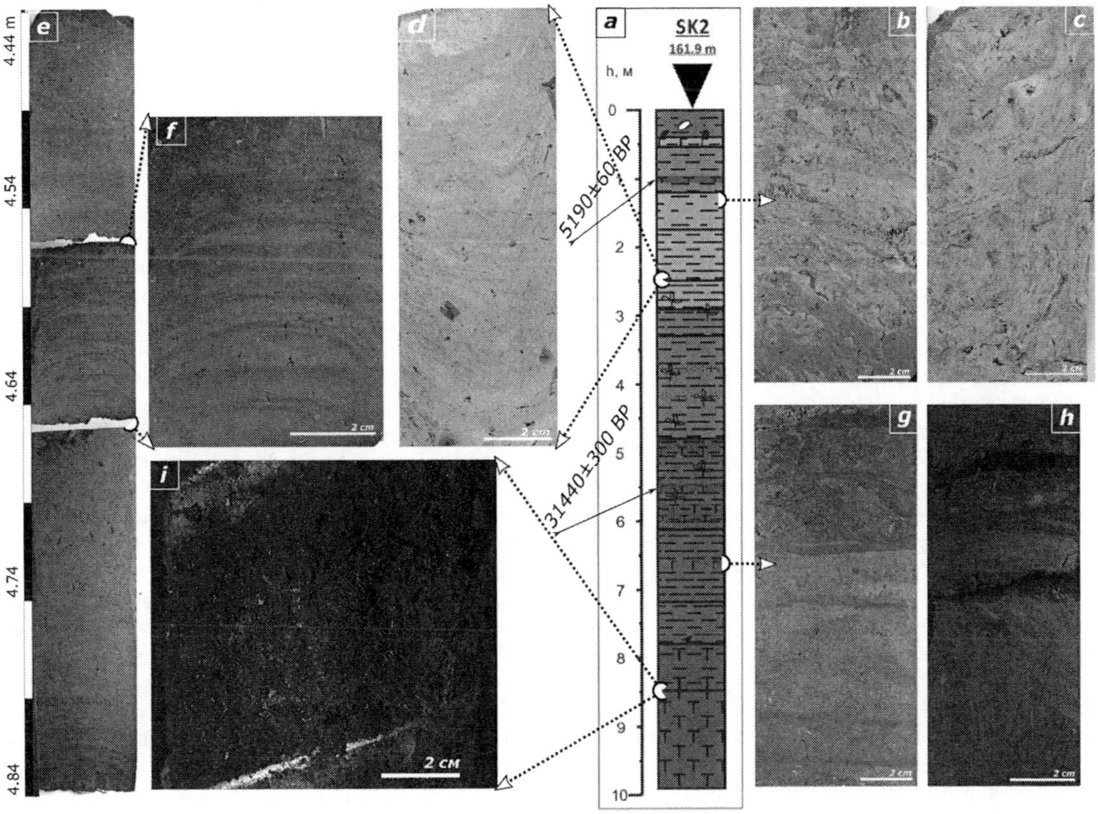

Figure 2. General lithostratigraphy and structure of postglacial sediments of the study site: **(a)** lithostrtigraphic column of SK2 core; **(b)** upper sedimentary layer – colluvium with irregular laminae-mottled structure and highly contrast texture; **(c)** fragment of cryogenic structure with subvertical laminae in colluvium; **(d)** regular cm-scale lamination in the upper lacustrine unit; slight cryogenic deformations, carbonate sand grains and ferrugineous nodules infilling fragments of plant detritus; **(e)** dm-scale stratification and **(f)** cm- and mm-scale laminae in the intermediate lacustrine unit; **(g)** cm-scale lamination with subangular clayey clasts and **(h)** fragments of dense peaty laminae in the lower lacustrine unit; **(i)** laminated basal peats.

fying in quantity downwards (Fig. 2). The lacustrine stratum splits into three distinctively different units. The upper approximately 1 m of light pale grey (with clearly visible mottling) clayey-silty loam without visible organic input (Fig. 2d) underlies the yellowish colluvial deposits with clear cryogenic structure evidencing slope wash, solifluction and cryoturbations (Fig. 2b–c), The lower >5 m of grey to dark-grey silty-clayey loam to clay with intermittent more sandy and gravelly layers (slope gully inputs) divide into intermediate and lower units. The former is most clearly fine-laminated (Fig. 2e–f), indicating deeper and calmer lake environments. The latter is dark-brown-grey with contrast stratification due to peat and plant detritus lenses 1–10 cm thick (Fig. 2g–h), representing changes of lake to peat bog sedimentation regime during dryer periods. Basal peats (Fig. 2i) likely correlate with the layers attributed to the Mikulino interglacial in adjacent locations (Rusakov et al., 2015). The ^{14}C date 31 440 ± 300 years BP (Ki-19671, Table 2) separating the lower and intermediate lacustrine units, recognizes that the former accumulated under milder conditions of the Early to Middle Valdai. A thin gyttja lens in overlying colluvium dates to 5190 ± 60 years BP (Ki-19676). This confines the deposition of the upper and intermediate lacustrine units to the second half of the Valdai. Particularly the upper fossilless unit was likely formed in cold Late Valdai periglacial environments with increasing role of slope processes input.

Lacustrine sedimentation was occasionally interrupted by increased fluvial activity represented by erosive contacts and lenses of sand with fine gravels (Z1, Z4, ChW1, ChW2) or completely ceased under the influence of incision-infill cycles (ORV). Radiocarbon ages of buried organic-rich lenses from lacustrine and overlying fluvial facies in adjacent cores Z1, Z4, Z5 at 173.8 m a.s.l. were acquired (Table 2). The results display lower lacustrine unit deposition succeeded by the secondary incision that likely occurred no later than the Middle Valdai (MIS-3) (Shishkina et al., 2019). Colluvial deposits and mantle loams at the top of the sedimentary sequences are typically ascribed to the Late Valdai periglacial period (MIS-2) (Velichko et al., 2004). However, pronounced organic-rich lenses (Table 2, dates 1 and 4) indicate that at least the upper 0.6 m formed under temperate conditions of the second half of the Holocene. Thus, two sets of ^{14}C dates obtained at different elevations (cores SK2 and Z-series) show good correspondence, suggesting regional scale of revealed lacustrine sedimentation trends.

Our study provides new evidence for an idea initially proposed by Rusakov and Sedov (2012) that observed surface drift layers traditionally interpreted as polygenetic cover loams (Semenov et al., 1972; Makeev et al., 2017) are more likely represented by lake deposits. The Late Pleistocene age of that lacustrine sequence is confirmed by its stratigraphic position: (1) overlying the eroded Middle Pleistocene till surface (MIS-6), at places with remnants of the Mikulino interglacial texturally-differentiated soils (Shishkina et al., 2019) or ephemeral peaty soils dated by U-Th method as MIS-5e (Rusakov et al., 2019); (2) occasionally underlain by peats interpreted as Early Mikulino by several spore-pollen analyses (Semenov et al., 1972) and U-Th dated as MIS-5 (Rusakov et al., 2019); (3) incorporating several organic-rich layers with newly obtained ^{14}C dates (Table 2) those are in accordance with the Middle Valdai ages reported by Rusakov et al. (2015, 2019). These humic layers probably reflect short-term pedogenesis on periodically dried lake bottoms and gully sides (Shishkina et al., 2019); (4) gradually disappearing plant macrofossils and organic carbon content in upper unit of the lacustrine sequence together with its upward change into typical mantle loams or colluvial deposits with clear cryogenic structures implicate scarce vegetation and severe periglacial conditions with increased cryogenic, slope wash and solifluction activity during the Late Valdai. In general, the observed lacustrine sedimentation can be considered prolonged and continuous through the majority of the Late Pleistocene with limited interruptions by dryer episodes or, at certain locations, fluvial incisions.

Certain lithological variability observed between cores and sections does not conceal the fact of their general similarities, including clear alternation of structural organization and textural composition of lacustrine strata reflecting the same variability of the lithodynamic (and possibly climatic) signals through time. Those could be considered as typical and most common sedimentary sequences of the Late Pleistocene lithogenesis of the eastern Borisoglebsk Upland. This observation contrasts with most of the published simplified models of polygenetic subaerial origin of mantle loams (Makeev et al., 2017). Lacustrine sediments of the Late Pleistocene age are present on all studied interfluve surfaces below 180 m a.s.l. and in infilled parts of the fluvial network. Their thickness is comparable with typical local relative elevations of the initial glacial topography that appears to be substantially smoothened if not entirely flattened by the lake infills. It means that the observed lacustrine sedimentation, together with cryogenic processes, as evident in several open sections (Belyaev et al., this issue), is the main agent of the Late Pleistocene landscape transformation in the study area.

More detailed genetic interpretation and absolute age evaluation of lacustrine deposits found on most of the lower interfluve surfaces and in gully side exposures is required. Two hypotheses can be advanced for further study. A long-term high water level stand of a jointed lake system inundating a series of depressions along the southern limit of the last Late Pleistocene ice cover from the Plecsheevo and Nero Lakes towards the Galich Lake may have occurred as proposed by Kvasov (1975) and Rusakov and Sedov (2012). However, the source of such recent waterbody of enormous volume is difficult to explain because meltwater generated during the Late Pleistocene glaciation should have been even less than that of the Middle Pleistocene glaciers. An alternative explanation may be an existence of a series of small local and relatively shallow bodies of standing water formed in numerous depressions of initial glacial topography (Antonov, 1998; Sheremetskaya et al., 2012) that periodically dried and were gradually filled by sediment from catchments slopes and peat formation. Even though this question has yet been left open, our data on grain size and organic content distribution and instrumental age determination do not support an existence of large-scale glacially induced lake. It rather speculates for the continuous lacustrine sedimentation in small isolated basins under significantly changing environmental conditions.

4 Conclusions

1. Detailed analysis of the eastern macroslope of the Borisoglebsk Upland suggests that most of the surface drift materials initially interpreted as polygenetic mantle loams were likely formed by continuous postglacial lacustrine deposition. Main arguments of such genetic interpretation are the dominance of fine-grained fractions; clear and texturally contrast multiscale lamination (up to five levels of laminae thickness); mottled colors typical for standing water environment. The Late Pleistocene age of that lacustrine sedimentation is stratigraphically evidenced by the Mikulino (Eemian) interglacial paleopedofeatures on the eroded surface of underlying Moscow glacial till and several organic-rich layers (paleosol humus horizons, peat, gyttja and plant debris lenses) in the lacustrine sequence dated to the Mikulino (MIS-5) interglacial and Middle Valdai (Middle Weichselian) interstadial (MIS-3).

2. Despite certain lithological variability observed between cores and sections, particularly those located on different elevations and in varying paleogeomorphic positions, general structural organization and textural composition of lacustrine stratum commonly ex-

hibits clearly evident alternation through time reflecting the same lithodynamic (and, possibly climatic) signals. Those can, therefore, be considered as typical and most common sedimentary sequences of the Late Pleistocene at the eastern Borisoglebsk Upland. This conclusion is certainly novel in contrast with most of the previously published models of dominantly subaerial polygenetic origin of mantle loams.

3. The observed lacustrine sediments of the Late Pleistocene age systematically occur both on all studied interfluve surfaces below 180 m and in buried parts of the fluvial network. Its thickness is comparable or, at places, even exceeds local relative elevations. It means that the lacustrine sedimentation was the main and yet underestimated agent of the Late Pleistocene topography transformation in the study area.

Data availability. For more information about the used data please contact the corresponding author Ekaterina V. Garankina (evgarankina@gmail.com).

Author contributions. All co-authors took part in field working and sampling. EVG, EDS and YVS carried out grain size and LOI analyses. IGS and PVA performed structural and palaeopedogenic analyses of both geological exposures and monoliths. VRB, EVG and IGS designed cartographic materials processing and artwork and prepared the manuscript with contributions from all co-authors.

Competing interests. The authors declare that they have no conflict of interest.

Special issue statement. This article is part of the special issue "Land use and climate change impacts on erosion and sediment transport". It is a result of the ICCE Symposium 2018 – Climate Change Impacts on Sediment Dynamics: Measurement, Modelling and Management, Moscow, Russia, 27–31 August 2018.

Acknowledgements. Special gratitude is to Andrey V. Dolgikh (IG RAS) for the ^{14}C dates calibration and a large number of the MSU students and staff, colleagues from the IG RAS and St Petersburg State University and friends who took part in fieldworks and helped with associated logistics.

Financial support. This research has been supported by the Russian Foundation for Basic Research (project 18-05-01118), the Department of Geomorphology and Paleogeography, Faculty of Geography, (GA no. AAAA-A16-11632810089-5) and the IG RAS (GA no. 0148-2019-0006).

References

ASTM D 2974-87: Standard Test Methods for Moisture, Ash, and Organic Matter of Peat and Other Organic Soils, USGA Green Section Record, March/April, 31–33, 1993.

Antonov, S. I.: Pleistocene lacustrine lithomorphogenesis at the southeastern Moscow region limits, in: History of Pleistocene lakes of the Eastern European Plain, Saint Petersburg, Nauka, 282–290, 1998 (in Russian).

Astakhov, V., Shkatova, V., Zastrozhnov, A., and Chuyko, M.: Glaciomorphological Map of the Russian Federation, Quaternary Int., 420, 4–14, 2016.

Belyaev, V. R., Semochkina, A. E., Streletskaya, I. D., Lisova, A. D., Garankina, E. V., Shorkunov, I. G., and Lugovoy, N. N.: Relic periglacial features and their impact on modern soil redistribution processes on cultivated slopes of the European Russia, PIAHS, submitted, 2019.

Garankina, E. V., Belyaev, V. R., Shorkunov, I. G., Andreev, P. V., Verlova, T. A., Shishkina, Yu. Vl., and Bondar, A. I.: Evolution of the Borisoglebsk Upland interfluves over the last 150000 years (Marginal zone of Moscow Glaciation, central part of Russian Plain), in: Practical Geography and XXI Century Challenges. International Geographical Union Thematic Conference, Moscow, Russia, 4–6 June 2018, Conf. Book, Part 1, 180–187, 2018.

Heiri, O., Lotter, A. F., and Lemcke, G.: Loss on ignition as a method for estimating organic and carbonate content in sediments: reproducibility and comparability of results, J. Paleolimnol., 25, 101–110, 2001.

Kleber, A. and Terhorst, B. (Eds.): Mid-Latitude Slope Deposits (Cover Beds), in: Developments in Sedimentology, 66, 1st Edn., Elsevier Science, 302 pp., 2013.

Konert, M. and Vandenberghe, J.: Comparison of laser grain size analysis with pipette and sieve analysis: a solution for the underestimation of the clay fraction, Sedimentology, 44, 523–535, 1997.

Kvasov, D. D.: Late Quaternary History of Large Lakes and Inland Seas of Eastern Europe, Leningrad, Nauka, 279 pp., 1975 (in Russian).

Lavrushin, Yu. A. and Chistyakova, I. A. (Eds.): Problems in the Stratigraphy of Quaternary Deposits and Paleogeography of the Yaroslavl Volga Basin: Proceedings of the International Symposium, GEOS-Press, Moscow, 2001 (in Russian).

Makeev, A., Kust, P., Lebedeva, M., Rusakov, A., Terhorst, B., and Yakusheva, T.: Soils in the bipartite sediments within the Moscow glacial limits of the Russian Plain: Sedimentary environment, pedogenesis, paleolandscape implication, Quatern. Int., 501, 147–173, https://doi.org/10.1016/j.quaint.2017.09.017, 2019.

Novskij, V. A.: Pleistocene of Yaroslavl Volga region, Moscow, Nauka, 236 pp., 1975 (in Russian).

Reinhardt-Imjela, C., Maerker, K., Schulte, A., and Kleber, A.: Implications of hydraulic anisotropy in periglacial cover beds for flood simulation in low mountain ranges (Ore Mountains, Germany), Die Erde, 149, 86–101, https://doi.org/10.12854/erde-2018-374, 2018.

Rusakov, A. and Sedov, S.: Late Quaternary pedogenesis in periglacial zone of northeastern Europe near ice margins since MIS 3: Timing, processes, and linkages to landscape evolution, Quaternary Int., 265, 126–141, 2012.

Rusakov, A., Nikonov, A., Savelieva, L., Simakova, A., Sedov, S., Maksimov, F., Kuznetsov, V., Savenko, V, Starikova, A., Korkka, M., and Titova, D.: Landscape evolution in the periglacial zone of Eastern Europe since MIS5: proxies from paleosols and sediments of the Cheremoshnik key site (Upper Volga, Russia), Quaternary Int., 365, 26–41, https://doi.org/10.1016/j.quaint.2014.09.029, 2015.

Rusakov, A., Sedov, S., Sheinkman, V., Dobrynin, D., Zinovyev E., Trofimova, S., Maximov, F., Kuznetsov, V., Korkka, M., and Levchenko, S.: Late Pleistocene paleosols in the extra-glacial regions of Northwestern Eurasia: Pedogenesis, post-pedogenic transformation, paleoenvironmental inferences, Quaternary Int., 501, 174–192, https://doi.org/10.1016/j.quaint.2018.03.020, 2019

Semenov, A. A, Tsukurova, A. M., Kvyatkovskaya, G. N., and Kuznetsov, V. K.: Geological map of Quaternary deposits, Scale 1 : 200000. Moscow series. Sheet O-37-XXVIII. Explanatory text, VSEGEI Publ., 1972 (in Russian).

Sheremetskaya, E. D., Borisova, O. K., and Panin, A. V.: Planation of Moscow glaciation periglacial zone during postglacial epoch, Geomorfologiya, 1, 92–106, 2012 (in Russian).

Shishkina, Yu. Vl., Garankina, E. V., Belyaev, V. R., Shorkunov, I. G., Andreev, P. V., Bondar, A. I., Potapova, V. I., and Verlova, T. A.: Postglacial Incision-Infill Cycles at the Borisoglebsk Upland: Correlations between Interfluve Headwaters and Fluvial Network, Int. Soil Water Conserv. Res., 7, 184–195, https://doi.org/10.1016/j.iswcr.2019.02.001, 2019.

Sudakova, N. G., Antonov, S. I., and Vvedenskaya, A. I.: Structure of glacial marginal zones in the center of the East European Plain, Bulletin of Moscow University, Series 5, Geography, 6, 55–61, 2013 (in Russian).

Velichko, A. A., Morozova, T. D., Nechaev, V. P., Rutter, N .W., Dlusskii, K. G., Little, E. C., Catto, N. R., Semenov, V. V., and Evans, M. E.: Loess/paleosol/cryogenic formation and structure near the northern limit of loess deposition, East European Plain, Russia, Quaternary Int., 152–153, 14–30, https://doi.org/10.1016/j.quaint.2005.12.003, 2006.

Proc. IAHS, 381, 21–24, 2019
https://doi.org/10.5194/piahs-381-21-2019

Contemporary gully erosion trend in the northern part of the forest-steppe zone of the Russian Plain: a case study from the Republic of Tatarstan, European Russia

Aidar Sharifullin[1], **Artur Gafurov**[1], **Regina Medvedeva**[1], **Valentin Golosov**[1,2,3], **Aleksandr Dvinskih**[1], **Artem Gusarov**[1], **and Benedict Essuman-Quainoo**[1]

[1]Institute of Environmental Sciences, Kazan Federal University, 18 Kremlyovskaya street, Kazan, 420008, Russia
[2]Research Laboratory for Soil Erosion and Fluvial Processes, Lomonosov Moscow State University, GSP-1, Leninskiye Gory street, Moscow, 119991, Russia
[3]Institute of Geography, Russian Academy of Sciences, Staromonetny per., 29, Moscow, 119017, Russia

Correspondence: Aidar Sharifullin (luleo123@mail.ru)

Published: 1 August 2019

Abstract. This article presents results for gully head cut retreat rates (GHRR) in the Pre-Kama region (Republic of Tatarstan, Russia) based on monitoring (1984–1994 and 2015–2018) and the interpretation of high-resolution satellite images (2004–2014). The results suggested that the mean linear GHRR have reduced by about five times, from 1.6 m per year (1983–1994) to 0.4 m per year (2015–2018). The main reason for the significant decrease in gully erosion was the increase in winter-spring air temperatures, which resulted in a reduction in the depth of soil freezing and surface runoff. The impact of land use, as well as erosion control measures, which were implemented in the study area during the late 1980s and early 1990s, was found to be less important.

1 Introduction

The problem of erosion of agricultural land in Russia has existed since the end of the 19th century (Dokuchayev, 1949; Kern, 1928). Detailed studies of gully erosion were undertaken in the mid-20th century by a number of researchers (Dedkov, 1990; Stupishin et al., 1984; Zorina, 2006, etc.) in the central and eastern parts of the Russian Plain. In the east of the Russian Plain, a map of gully density based on topographic maps of different scales (1 : 100 000, 1 : 50 000 and 1 : 25 000) was constructed for the period covering 1950–1970s. The first data for GHRR were obtained in 1953, based on the interpretation of aerial photographs. Unfortunately, continuous gully headcut retreat monitoring that was stopped in 1994.

In recent decades, the European territory of Russia (ETR) has seen both climate change and a change in land use associated with the collapse of the Soviet Union in 1991 (Apukhtin and Kumani, 2012; Ivanov et al., 2017, etc.). These changes could be expected to affect the intensity of gully erosion. However, the absence of modern gully monitoring in the north of the forest-steppe zone does not provide a basis for drawing conclusions and, therefore, it is necessary to resume analysis. The purpose of this study is to provide a quantitative assessment of modern gully erosion rate on the arable lands of the Pre-Kama region of the Republic of Tatarstan, European Russia.

2 Materials and methods

The research area is located in the western part of the Pre-Kama region (16 620 km^2) of the Republic of Tatarstan (Fig. 1), between the Volga, Kama and Vyatka Rivers. The average altitude within the study area is 150 m. Gully density ranges from 0.1 to 0.5 $km\,km^{-2}$ (Golosov et al., 2018). The geological structure of the territory of the Western Pre-Kama region is composed of Permian rocks (mainly limestones and dolomites) and Quaternary sediments (mainly aleurites and clays). The climate of the study area is continental, with warm summers and moderately cold winters.

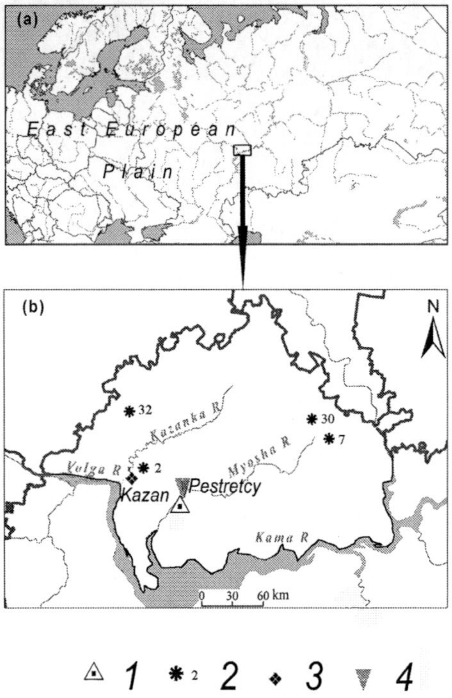

Figure 1. The location of the Pre-Kama region in the Russian Plain **(a)** and the Republic of Tatarstan **(b)**. (Legend: 1 – the study region, 2 – the study sites with measurement of GHRR, based on the interpretation of high-resolution satellite images, 3 – the study sites with measurement of GHRR, based on a monitoring, 4 – the Kazan-Opornaya meteorological station, 5 – the Pestretcy gauging station (at the Myosha River).)

According to the Kazan weather station, the average annual temperature is +4.1 °C (1950–2017). The warmest month is July (+19.8 °C), and the coldest month is January (−12 °C). The annual average rainfall is 557 mm, of which 69 % falls in the warm season.

Gully erosion was investigated using two approaches: monitoring and interpretation of high-resolution satellite images. The headcuts of 48 gullies, located within four small catchments, were monitored from 1983 to 1994. The monitoring stopped in 1995 and was re-launched in 2015. GHRR was determined once per year for 1983–1994 and twice per year (in May and September) for 2015–2018 by measuring the distance from the gully head to a fixed metal steak, placed upslope from gully head.

The interpretation of different-time satellite images was carried out for 2004 and 2014. Images were, taken from the WorldView-2, WorldView-3 and GeoEye satellites. The input images were first processed and standardized to a common scale and projection. In the next stage, thalwegs were digitized from images maded in 2004 and 2014 for each gully. Finally, the maps were overlain and the thalwegs used to calculate GHRR.

3 Results

Mean linear GHRR for the period 1983–1994 was estimated at $1.6 \pm 0.5\,\mathrm{m\,yr^{-1}}$. During this time interval the highest GHRR were atypically associated with bottom gullies growth. For the period 2015–2018, the GHRR was estimated at $0.4 \pm 0.1\,\mathrm{m\,yr^{-1}}$. A decrease in GHRR was therefore revealed for 2015–2018 compared to 1983–1994.

The linear GHRR for slope gullies according to satellite imagery were ranged from 0 to $3.4\,\mathrm{m\,yr^{-1}}$ during 2004–2014, with an average rate of $0.3\,\mathrm{m\,yr^{-1}}$. At the same time, about 70 % of all the gullies were found to exhibit inactive headcuts.

Previous studies conducted in the territory of the Western Pre-Kama have reported a slow attenuation in the growth rate of the gullies: $2\,\mathrm{m\,yr^{-1}}$ in 1953–1958 (Stupishin et al., 1980), $0.8\,\mathrm{m\,yr^{-1}}$ in 1958–1985 (Butakov et al., 1987), $1.6\,\mathrm{m\,yr^{-1}}$ in 1983–1994, $0.3\,\mathrm{m\,yr^{-1}}$ in 2004–2014, and $0.4\,\mathrm{m\,yr^{-1}}$ in 2015–2018.

On the East European Plain, about 80 % of the increase in gullies is observed during the period of spring snowmelt. Therefore, meteorological parameters characterizing winter-spring conditions (Table 1) need to be taken into account when interpreting data for GHRR. All of the parameters that were taken into account in this study are characterized by significant changes, especially over the recent decades. Thus, since the 1980s, there has been an increase in air temperature during the winter spring periods. Also, there has been an increase in precipitation and water storage in snow. However, the increase in precipitation almost did not affect the height of the snow cover. The observed changes in such winter-spring meteorological parameters have generally good agreement with water flow characteristics in the Myosha River during the spring flood (Fig. 2). Thus, since the 1980s, maximum annual discharges and total water runoff depth in the basin during the spring have been decreased by 31 %. Therefore, it is very likely that the reduction of surface water runoff was the main reason for decreasing of the GHRR in the region.

One of the most significant human factors that influences gully erosion rates, by changing the structure of the upper soil layer (e.g. Valentin et al., 2004), is the area of cultivated land. However, there were no significant changes in the area of cultivated land in the Myosha River basin in 2015 compared to 1985 (Ivanov et al., 2017). The study area was actively ploughed even in the 1990s. The main crops grown are wheat, barley and corn. In the recent years, perennial grasses have also been introduced into rotations.

Another reason for the reduction in the GHRR in this region was the implementation of erosion control measures that began in the late 1970s. In total, about 120 erosion control

Table 1. Some climatic characteristics for the study region T_{aw} – the mean long-term air temperature for the winter period (December to February), T_{am} – the mean long-term air temperature for March, N – the number of days in the winter period with daily temperatures above 0 °C, A – the duration of the cold period (the number of days with a stable average daily temperature below zero), P – the mean long-term precipitation for the cold period, H – the mean long-term snow depth for the winter period, S – the average long-term water storage in the snow for the first decade of March.

Periods	T_{aw}, °C	T_{am}, °C	N, d	A, d	P, mm	H, cm	S, mm
1980–1989	−10.1	−5.6	2.6	154	167	29.1	96
1990–1999	−9.8	−3.9	4.5	150	185	29.9	–
2000–2009	−8.7	−2.7	7.7	138	200	29.0	134
2010–2018	−9.6	−3.9	6	138	228	40.1	141

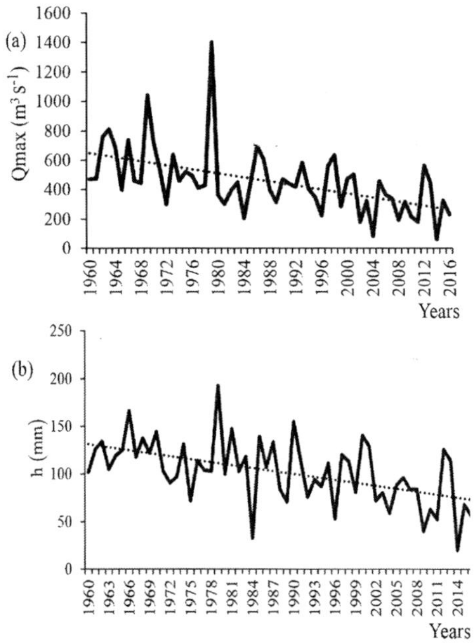

Figure 2. Long-term changes in water flow recorded at the Myosha River, **(a)** maximum annual water discharges (during the spring snowmelt period), **(b)** water runoff depth in the basin during the spring flood period.

works were carried out within the study area. But, in the absence of the regular maintenance of anti-erosion structures , their effectiveness reduced considerably. Hereby meaning the reduction in GHRR is more associated with changing meteorological conditions rather than with application of erosion control measures.

4 Conclusions

A decrease in the GHRR from 1.6 m yr^{-1} (1983–1994) to 0.4 m yr^{-1} (2015–2018) was established for the Western Pre-Kama region of the Republic of Tatarstan (European Russia), on the basis of monitoring and the interpretation of satellite images. The linear GHRR derived from the monitoring were found to be in good agreement with an estimate (0.3 m yr^{-1} for 2004–2014) based on the interpretation of satellite images. The change of hydrometeorological conditions, especially winter air temperatures is the main reason for the decrease of GHRR. Land use and erosion control measures did not affect the development of gullying processes. The recent trend of a decline in gully erosion intensity was consistent with the earlier-reported decrease in soil erosion in the study area (Sharifullin et al., 2018).

Data availability. We used a database of gullies, which is part of the big database. I've added its in the text (Vanmaercke et al., 2016) and in the references. Part of the data of gullies was resently received during fieldwork and was not been published In addition, meteorological information about daily precipitation data were provided by the RIHMI World Data Center (http://aisori-m.meteo.ru/waisori/, RIHMI World Data Center, 2018). Information about hydrological characteristic was downloaded from Automated information system of the state monitoring of water bodies (AIS SMWB) (https://gmvo.skniivh.ru/, AIS SMWB, 2018) and were collected from Hydrological Yearbooks, which are available-only in printed version.

Author contributions. VG. conceived the idea of gully erosion assessment and correct the text. AS wrote the manuscript with support by BE and AG. RM and AD collected and analysed the data of gully erosion.

Competing interests. The authors declare that they have no conflict of interest.

Special issue statement. This article is part of the special issue "Land use and climate change impacts on erosion and sediment transport". It is a result of the ICCE Symposium 2018 – Climate Change Impacts on Sediment Dynamics: Measurement, Modelling and Management, Moscow, Russia, 27–31 August 2018.

Acknowledgements. This work was funded by the Russian Foundation for Basic Research, project no. 18-35-00326 (fieldwork on measuring GHRR), and Russian Science Foundation, project no. 19-17-00064 (data collection on hydrological and meteorological parameters).

Financial support. This research has been supported by the Russian Foundation for Basic Research (project no. 18-35-00326) and Russian Science Foundation (project no. 19-17-00064).

References

Apukhtin, A. V. and Kumani, M. V.: Recent changes in the conditions of spring floods of rivers in Kursk region, Elec. Sci. Jour. of the Kursk. St. Univ., 1, 300–311, 2012 (in Russian).

AIS SMWB (Automated information system of state monitoring of water bodies): Hydrological characteristic, https://gmvo.skniivh.ru/, last accessed 10 November 2018.

Butakov, G. P., Dvinskih, A. P., Nazarov, N. N., and Rysin, I. I.: Modern gully erosion in the east of the Russian Plain, Geomorfologiya, 2, 43–46, 1987.

Dedkov, A. P. (Ed.): Ovrazhnaya eroziya vostoka Russkoy ravniny (Gully Erosion in the East of the Russian Plain), Kazan: Kazan State University (Publ.), 142 pp., 1990 (in Russian).

Dokuchayev, V. V.: Izbrannye trudy (Selected papers), edited by: Polynova, B., Moscow: Publishing House of the Academy of Sciences of the USSR, 223 pp., 1949 (in Russian).

Golosov, V., Yermolaev, O., Rysin, I., Vanmaercke, M., Medvedeva, R., and Zaytseva, M.: Mapping and spatial-temporal assessment of gully density in the Middle Volga region, Russia, Earth Surf. Proc. Land., 43, 2818–2834, 2018.

Ivanov, M. A., Prishchepov, A. V., Golosov, V. N., Zalyaliev, R. R., Efimov, K. V., Kondrat'eva, A. A., Kinyashova, A. D., and Ionova, Yu. K.: Changes of cropland area in the river basins of the European part of Russia for the period 1985–2015, as a factor of soil erosion dynamics, Sovremennye Problemy Distantsionnogo Zondirovaniya Zemli iz Kosmosa, 14, 149–157, 2017.

Kern, E. E.: Ovragi, ikh zakrepleniye, obleseniye i zapruzhivaniye (The gullies, its consolidation, afforestation and damming), Moscow: State publishing house, 163 pp., 1928 (in Russian).

RIHMI World Data Center (Russian Research Institute of Hydrometeorological Information – World Data Center Hydrometeorology data service system): meteorological information about daily precipitation data, http://aisori-m.meteo.ru/waisori/, last accessed: 12 December 2018.

Sharifullin, A. G., Gusarov, A. V., and Golosov, V. N.: Assessment of contemporary erosion/sedimentation trend within a small cultivated catchment in the Republic of Tatarstan (European Russia), Geomorfologiya, 3, 93–108, 2018.

Stupishin, A. V., Duglav, V. A., and Lapteva, N. N.: Geograficheskiy analiz ovrazhno-balochnykh sistem v predelakh Tatarskoy ÀSSR (Geographical analysis of gully/dry valley systems in the Tatar ASSR), Kazan: Kazan State University, 152 pp., 1980 (in Russian).

Valentin, C., Rajot, J. L., and Mitja, D.: Responses of soil crusting, runoff and erosion to fallowing in the sub-humid and semi-arid regions of West Africa, Agr. Ecosyst. Environ., 104, 287–302, 2004.

Vanmaercke, M., Poesen, J., Mele, B. V., Demuzere, M., Bruynseels, A., Golosov, V., Bezerra, J. F.-R, Bolysov, S., Dvinskih. A., Frankl, A., Fuseina, Yu., Guerra, A. J.-T., Haregeweyn, N., Ionita, I., Imwangana, F. M., Moeyersons, J., Moshe, I., Samani, A. N., Niacsu, L., Nyssen, J., Otsuki, Yo., Radoane, M., Rysin, I., Ryzhov, Yu. V., and Yermolaev, O.: How fast do gully headcuts retreat?, Earth-Sci. Rev., 154, 336–355, https://doi.org/10.1016/j.earscirev.2016.01.009, 2016.

Zorina, E. F.: Geografiya ovrazhnoy erozii (Geography of gully erosion), Moscow: MSU, 324 pp., 2006 (in Russian).

Proc. IAHS, 381, 25–29, 2019
https://doi.org/10.5194/piahs-381-25-2019

Modern patterns of spatial distribution of the main soil parameters: case study of the Republic of Tatarstan, Russia

Artur Gafurov[1,2] **and Aidar Sharifullin**[2]

[1]OpenLab Cosmoecology, Kazan Federal University, Kazan, 420008, Russia
[2]Department of Landscape Ecology, Kazan Federal University, Kazan, 420008, Russia

Correspondence: Artur Gafurov (gafurov.kfu@gmail.com)

Published: 1 August 2019

Abstract. The territory of the Republic of Tatarstan (RT) belongs to the region with the most developed agriculture in Russia and the Middle Volga region. However, the formation of a system of accurate farming in the Republic is just beginning. To ensure precision farming, it is necessary to precisely know cropland borders, as well as the agrochemical properties of soils (pH, humus content, potassium and phosphorus content). The present study was carried out based on aerial images analyses of the most intensively used part of the RT both with identification of soil indicators for each of the fields. The availability of information on the spatial location of various cultivated crops makes it possible to assess the role of soil characteristics in these crops.

1 Introduction

Contemporary agricultural sector is aimed at reducing the cost per unit of output and increasing productivity per unit of expanded resources. "Precision agriculture" or "precision farming" described a method of agricultural monitoring which included the use of frequently updated information gathered by remote sensing satellites. Instead of plowing, sowing, and fertilizing "by eye", as was done throughout the previous history of agriculture, today farmers can accurately calculate the number of seeds, fertilizers and other resources for each plot of the field with an 1 m accuracy. Based on precise field map which is done through satellite and laboratory data analyses, the characteristics of each plot are identified. This is efficient way to resources use. Thus, it is possible to avoid overspending of resources where they were previously used in excess, and to increase the productivity of those parts of the field that were previously undersupplied in fertilizers, plowing or irrigation. The system of precision farming is not a strictly defined set of techniques and technical means, but rather a general concept based on the use of satellite positioning technologies (GPS), geographic information systems (GIS), accurate mapping of fields, etc. Studying soil characteristics, such as acidity, content of minerals and humus is an important task of modern agriculture. Understanding of spatial laws of these characteristics are particularly important. Without knowledge of these laws, precision farming is impossible. The aim of the work is to study the spatial regularities of the main soil parameters on the example of the agrarian developed part of the Republic of Tatarstan according to remote sensing data of the Earth. This requires solving the following tasks: collecting data on the main soil characteristics under field conditions, determining the boundaries of fields on the territory of agricultural lands of the Republic, determining the type of vegetation on the selected fields, searching for a connection between soil characteristics and natural anthropogenic factors, analyzing the identified patterns.

2 Study area

Tatarstan is located in the centre of the Russian Federation on the East European Plain, at the confluence of the two largest rivers – the Volga and Kama. The territory of the Republic is mostly laying within forest-steppe zone with small elevations on the right bank of the Volga and the south-east of the Republic. 90 % of the territory lies at an altitude of no more than 200 m above sea level (Yermolaev et al., 2018b). More than

Published by Copernicus Publications on behalf of the International Association of Hydrological Sciences.

16 % of the territory of the Republic is covered with forests consisting of mainly deciduous trees (oak, linden, birch, aspen), conifers are represented by pine and spruce. The climate is moderately continental. The soils are distinguished by a wide variety – from gray forest and podzolic in the north and west to various types of black soil in the south of the Republic (32 % of the area). Almost all the territory of the Republic is intensively used in agriculture (Usmanov et al., 2015). Crops such as peas, wheat, rye, barley, maize, oats, sunflower, colza, sugar beets and herbs are grown here.

One of the most important elements of soil fertility is their agrochemical characteristics. As a result of the stable land leaching, the area of acidic soils compared to the previous survey cycle (7–8 years) decreased by 160 thousand hectares and in 2007 was 1361.6 thousand hectares, or 39.2 % of the arable land, there are 1090.4 thousand hectares – slightly acid, 241.3 thousand hectares – medium acid and 29.9 thousand hectares - strongly acid. The highest level of acidic soils in the Agryz district is 62.9 %, Aksubaevsky – 65.9 %, Alkeevsky – 69.4 %, Tukayevsky – 64 %, and Rybno-Slobodsky – 72.9 %. Currently, the weighted average pH value in the Republic is 5.6.

The progress in acidic soils reduction is limited by the ongoing use of physiologically acidic mineral fertilizers, high consumption of calcium and magnesium carbonates due to removal with the crop, as well as soil removal due to erosion.

3 Material and methods

To implement the system of precision farming on more than 3000 fields used in agriculture, the delineation of the boundaries based on aerial survey data was done. Aerial photography was carried out with 70 % longitudinal and transverse overlapping of images, and the coordinates of the centres of the image were determined using of the GNSS receiver (Gafurov, 2018). The reference error was within 0.3 m in height. The resolution of orthophotos obtained was 0.25. Field work was carried out to determine the soil properties, such as acidity, humus, the content of potassium and phosphorus oxides in each field. These data were combined into a database based on which some statistical indicators were obtained (Table 1).

Type of vegetation for more than 1000 fields was determined. These fields, in the future, were used as etalons. The classification was carried out according to satellite imagery of the Sentinel-2 system according to the methodology given in article (Ivanov et al., 2017). The obtained data on the type of cultivated crop were used to analyze the dependence of soil indicators on the type of vegetation.

The classification of vegetation types was carried out using QGIS and EnMap plugin. ANOVA and MANOVA was conducted in the R language environment.

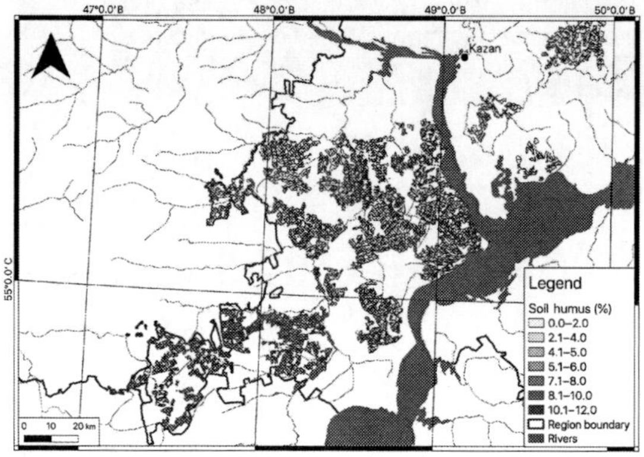

Figure 1. Distribution of humus content of soil on the example of west part of RT.

4 Results and discussion

Because of the research, it was found that the spatial distribution of the main plant nutrients in the study area is uneven. The obtained results indicate certain regularities in the distribution of soil fertility characteristics (Fig. 1). The nature of the distribution often depends on the topography and other physiographic conditions, but is also associated with the agrogenic load on the landscape. Around large settlements, where soils are used as arable land very intensively, there is a lack of basic plant nutrients. For the relationship between the content in the soil of nutrients for plants and the characteristics of the relief was carried out analysis of variance (ANOVA). Heights and slopes in degrees were used as an independent variable.

The results obtained suggest that there is a strong relationship between the height of the relief and pH values, the content of potassium and phosphorus in the soil ($p<0.001$). Also, a strong dependence was found between the values of average slopes on the field and pH values, the content of potassium and humus in the soil ($p<0.001$) (Fig. 2). With an increase in the height of the relief and slope, the acidity of the soil decreases to neutral and higher, which relates to an increase in soil washout with a steeper relief. Considering the chalky limestone deposits as a bedrock, this leads to a general leaching of the arable layer. The high influence of soil erosion on soil fertility is also confirmed by the fact that with a decrease in surface slopes the content of humus in the soil increases, which, again, is associated with a decrease in soil washability and accumulation of humus in the soil. Indicators of humus content decreases as it approaches the lower part of the river network, which indirectly may indicate low soil protection from surface runoff. A similar results are observed for the content of potassium and phosphorus oxides in the soil and are according to previous studies (Yermolaev et al., 2018a).

Table 1. Average values of soil indicators for different parts of the Republic of Tatarstan.

Soil indicator	Humus (%)	pH	K_2O ($mg\,kg^{-1}$)	P_2O_5 ($mg\,kg^{-1}$)
West part (2443*)	2.1–4.0	5.1–5.5	121–170	151–250
North part (717*)	2.1–4.0	5.6–6.0	81–120	151–250
South part (96*)	6.1–8.0	5.1–5.5	81–120	101–150
Mean (3256*)	2.1–4.0	5.1–5.5	121–170	151–250

* Number of fields

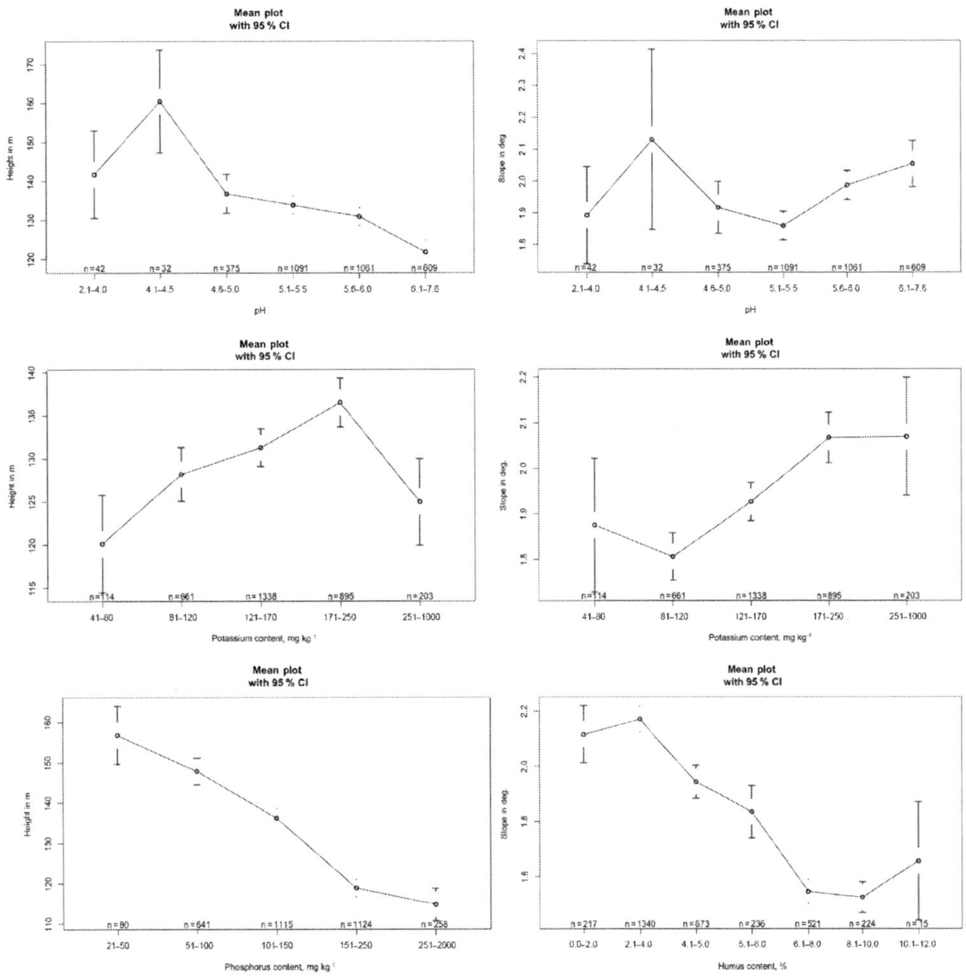

Figure 2. The ANOVA results between the content in the soil of nutrients for plants and the characteristics of the relief.

The analysis of the dependence of soil indicators on soil types showed spatial regularities characteristic of the main soil types: with the movement to the south, the soil acidity changes in the direction of its decrease, and the humus content index in the soil increases (Fig. 3). Indicators related to fertilizer application, such as the content of phosphorus and potassium in the soil, either have a weak connection with the type of soil, or have no significant relationship at all. This fact shows the supporting nature of fertilization without considering the natural features of the territory.

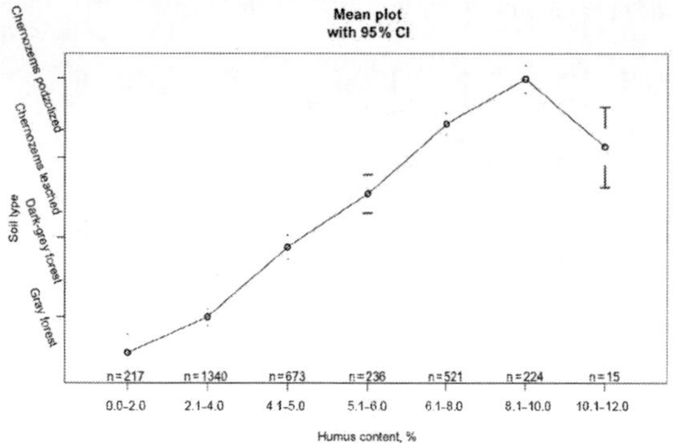

Figure 3. The ANOVA results between the soil type and the humus content.

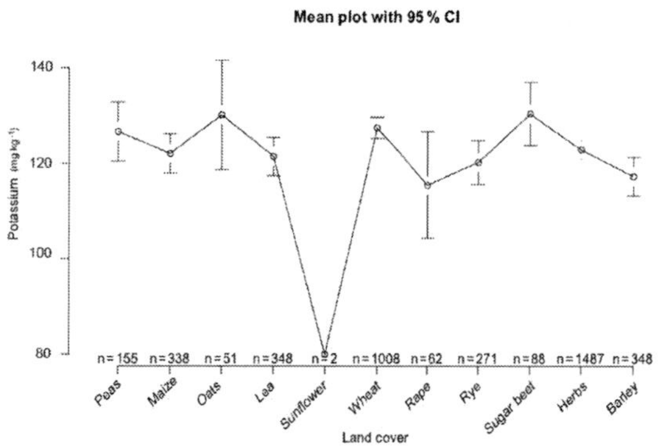

Figure 4. The ANOVA results between the culture and the potassium content in the soil.

If we consider the dependence of the content of various elements relative to each other, then no significant correlation coefficients were found.

The ANOVA (Fig. 4) and MANOVA showed a significant dependence ($p<0.05$) of the cultivated crop on the potassium content in the soil, as well as the combined content of potassium and phosphorus. No significant correlation was found between the soil acidity index, the humus content and the type of cultivated culture.

This indicates that fertilization in the soil is not related to the yield and fertility indicators of the soil. Fertilization in this case is carried out according to the recommendations on fertilizer application from the Soviet period. Alkaline fertilizer application is also weakly related to soil acidity indicators. All this indicates the absence of so-called precision farming in most of the agricultural area of the Republic.

5 Conclusions

The results obtained during work suggest that precision farming in the Republic of Tatarstan is not applied. The content of nutrients in the soil is reduced due to soil erosion. Fertilizers applied to the soil are only of a supportive nature and are not related to the type of growing crop.

Data availability. Due to the fact that the data represent a commercial secret, we cannot provide public access to the results of soil analysis. The provision of data will be decided individually upon the request of interested scientists to the correspondence author.

Author contributions. The authors carried out work on field surveys with UAVs, analysis of land cover types and statistical analysis (AG), as well as collecting and analyzing soil samples in all fields and collecting information about the study area (AS).

Competing interests. The authors declare that they have no conflict of interest.

Special issue statement. This article is part of the special issue "Land use and climate change impacts on erosion and sediment transport". It is a result of the ICCE Symposium 2018 – Climate Change Impacts on Sediment Dynamics: Measurement, Modelling and Management, Moscow, Russia, 27–31 August 2018.

Acknowledgements. The study was carried out thanks to the Russian Science Foundation (collection of data on vegetation on the fields and UAV survey) and the Russian Foundation for Basic Research (collection of data on soil indicators).

Financial support. This research has been supported by the Russian Science Foundation (grant no. 19-17-00064) and the Russian Foundation for Basic Research (grant no. 18-35-00326).

References

Gafurov, A. M.: Small catchments DEM creation using Unmanned Aerial Vehicles, IOP Conf. Ser.: Earth and Env. Sci., 107, 12005, https://doi.org/10.1088/1755-1315/107/1/012005, 2018.

Ivanov, M. A., Prischepov, A. V., Golosov, V. N., Zalyaliev, R. R., Yefimov, K. V., Kondratieva, A. A., Kinyashova, A. D., and Ionova, Yu. K.: Method of croplands dynamics mapping in river basins of the European part of Russia for the period of 1985–2015, Sovremennye Problemy Distantsionnogo Zondirovaniya Zemli iz Kosmosa, 14, 161–171, https://doi.org/10.21046/2070-7401-2017-14-5-161-171, 2017.

Usmanov, B., Yermolaev, O., and Gafurov, A.: Estimates of slope erosion intensity utilizing terrestrial laser scanning, Proc. IAHS, 367, 59–65, https://doi.org/10.5194/piahs-367-59-2015, 2015.

Yermolaev, O. P., Gafurov, A. M., and Usmanov, B. M.: Evaluation of Erosion Intensity and Dynamics Using Terrestrial Laser Scanning, Eurasian Soil Sci., 51, 814–826, https://doi.org/10.1134/s1064229318070037, 2018a.

Yermolaev, O. P., Mukharamova, S. S., Maltsev, K. A., Ivanov, M. A., Ermolaeva, P. O., Gayazov, A. I., Mozzherin, V. V., Kharchenko, S. V., Marinina, O. A., and Lisetskii, F. N.: Geographic Information System and Geoportal "River basins of the European Russia", IOP Conf. Ser.: Earth Environ. Sci., 107, 012108, https://doi.org/10.1088/1755-1315/107/1/012108, 2018b.

Proc. IAHS, 381, 31–35, 2019
https://doi.org/10.5194/piahs-381-31-2019

Impacts of fluvial processes on medieval settlement Lukovskoe (Tatarstan, Russia)

Artur Gafurov[1], Iskander Gainullin[2], Bulat Usmanov[1,3], Petr Khomyakov[1], and Alexey Kasimov[3]

[1]Institute of Environmental Sciences, Kazan Federal University, Kazan, 420008, Russia
[2]Institute of archaeology. A. H. Khalikov, Academy of Sciences of the Republic of Tatarstan, Kazan, 420012, Russia
[3]SRL "Interdisciplinary Innovative and Scientific-Practical Archaeological and Ethnological Research", Kazan Federal University, Kazan, 420008, Russia

Correspondence: Artur Gafurov (gafurov.kfu@gmail.com)

Published: 1 August 2019

Abstract. This work is continuation of the research aimed at developing of a system for analysing of risks of destruction of archaeological objects of Volga-Bulgaria period (X–XIII centuries AD) on the territory of the Republic of Tatarstan. Most of the Volga-Bulgaria settlements located on the small rivers banks. This fact is determining the risk of their destruction by natural processes. In order to identify the risk of transformation of archaeological sites by fluvial processes bank erosion dynamics at Lukovskoe fortified settlement evaluated. Modern field survey with use of UAV and GNSS methods applied to study the relief of monument territory. The measured data were analysed using the Geographical Information System (GIS) to evaluate intensity of bank erosion. Historical maps, archival remote sensing data and actual orthophotoplans compared to get quantitative characteristics of monument territory damage. The main factors that influence bank erosion at Lukovskoe fortified settlement placement are the meandering of the Kubnya riverbed, bank height and the ground that forms it. This study shows that remote sensing and 3-D-modelling on the base of UAV survey is very important in study of fluvial processes dynamics as a factor of archaeological objects destruction. The results of research will help to identify trends in monuments state and to quantify the risks of their destruction.

1 Introduction

The state of archaeological sites consists of interacting natural and anthropogenic components that form under the influence of human activity and natural processes over time. To characterize the effects of natural processes on the current state of medieval sites, the most recent interdisciplinary methods and technologies are required (Gainullin et al., 2017). Accordingly, the development of cultural heritage preservation strategies based on advanced methods is an essential part of modern archaeological research (Nicu, 2018).

Remote sensing, geo-informatics and field survey methods can be used to evaluate spatial and temporal changes in landscapes and related historical sites. These analytical methods provide quantitative assessments of impact of natural hazards and the risks of related processes on archaeological sites.

Notably, multitemporal aerial and satellite imagery are widely used in the preservation assessment of historical and cultural heritage objects (Gainullin et al., 2017). Increasingly, unmanned aerial vehicles (UAV) are being used to conduct remote surveys of local archaeological objects and create 3-D-models (Fernandez-Hernandez et al., 2015). The use of high-precision geodetic equipment is common and essential for the study of archaeological sites. For example, geodesic methods, such as GNSS technology (Gafurov, 2018), topographic survey with total station (Usmanov et al., 2018) and laser scanning equipment (Usmanov et al., 2015; Yermolaev et al., 2018) have been used to evaluate the dynamics and effects of exogenous processes that threaten monuments. Thus, the use of modern technologies provides accurate data collection, help to build complex multi-layer maps and form archaeological objects geodatabases.

Published by Copernicus Publications on behalf of the International Association of Hydrological Sciences.

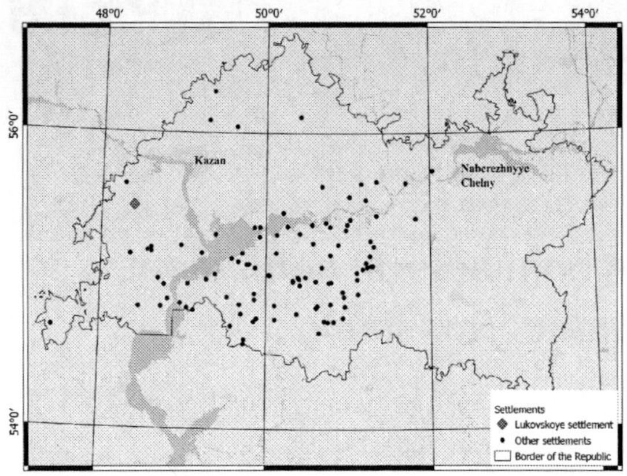

Figure 1. Study area.

In 2017, 2751 objects of cultural heritage (historical and cultural monuments) were under the influence of natural and anthropogenic factors in Republic of Tatarstan (RT), Russia. Accordingly, the loss of such artefacts will have implications for the historical legacy of the Tatarstan people. The last detailed study of archaeological monuments of Tatarstan was conducted in the 1970s and more recent investigation of these archaeological heritage sites of significance is necessary.

The Republic of Tatarstan is located in the eastern part of the East European Plain, near the confluence of Volga and Kama rivers (Fig. 1) and covers, an area of 68 000 km^2. The average elevation is 170 m above sea level but some areas rise to 300–350 m. The study area is located on the east European Plain that rises from the Volga River to the spurs of the Ural Mountains. Erosional features in the landscape formed by flowing water include gullies and ravines that dissect the slopes of uplands and river valleys. Four main large rivers (Volga, Kama, Belaya and Vyatka) and an additional 500 small rivers over 10 km in length transect this region. Many of the prominent archaeological sites in this region are located on the banks and terraces of small rivers (Yermolaev et al., 2012).

The goal of this study is to develop a system to evaluate the potential risk of fluvial processes of impacting archaeological objects of Volga-Bulgaria that were constructed for the period (X–XIII centuries AD) in Tatarstan. Here, defensive fortifications (shafts, ditches), are examined because they are easily identified using aerial survey data. The present study reports the effects of riverbed dynamics on a Lukovskoe fortified settlement.

2 Materials and methods

An evaluation of riverbed dynamics in this study was evaluated using GIS and remote sensing techniques described by Ivanov et al. (2017). In addition, historical maps, topographical maps, aerial images (Soviet aerial images from 1958 and 1980) and field surveys with UAV and GNSS were used to other important information which was used in an integrated GIS database. For a better understanding of the fortification system, river dynamics and negative exogenous processes, a 3-D model was developed using UAV photos made during fieldwork in 2017 and 2018. The model was created using AgiSoft Photoscan software. The procedure involved the following stages:

1. Gathering the information – dedicated literature, cartographic sources, archive data, RS data, aerial and satellite photographs, on-site photographs;

2. Field surveys – specification of location and visual features of settlement; photo identification of current monument state; GNSS survey of reference points; aerial survey of archaeological object under study.

   Aerial photographs of the area were produced with a multi-rotor UAV DJI Phantom 4 using a 12-megapixel camera mounted on a quadrocopter. A smartphone using DronDeploy software Aerial photography was performed with the following parameters: – height – 50–100 m; – pictures overlapping – 60 %–80 %; – camera position – 90°; – coverage – area survey; – meteorological conditions – no precipitation, wind no more than $15\,m\,s^{-1}$.

   Coordinates in plan and height during aerial survey were determined using GPS built in the UAV. This accuracy (4–5 m) is not sufficient for the construction of a topographic plan, digital terrain model (DTM) and digital elevation model (DEM) of an archaeological heritage object. To increase the accuracy of the survey before the flight, a network of control points was identified banner fabric with holes in the corners to fix geodetic pole with antenna. For high precision coordinates and height detection, a double frequency GNSS receiver with external antenna and GCPs centres were used. The survey was performed in Real Time Kinematics (RTK) regime in World Geodetic System 84 (WGS84) with real-time corrections by satellite reference stations. Coordinates of the GCPs centers improved UAV data processing accuracy.

3. Office analysis – all the maps and aerial photos were georeferenced to identify the most dynamic areas prone to river erosion, to identify the stages of archaeological heritage destruction, and 3-D modelling.

   Photogrammetric works were performed in Agisoft Photoscan in the following sequence: – images inspection; – pictures alignment; – cloud of points construction; – digital elevation model and orthophoto construction.

   DEM analysis, mapping and calculations were conducted in Golden Software Surfer 13 software. A digital terrain model (DTM) with a step of 0.5 m showing the altitude characteristics of the settlement territory was generated by the point cloud. Profiles, inclination and aspect maps described the morphometric characteristics of settlement relief. Ultra-high resolution orthophotos with a pixel size of 5 cm permitted creation of electronic layers for topographic plans to be constructed in MapInfo Professional 15.0 software.

3 Results and discussion

Lukovskoe (Yapanchino) fortified settlement (Fig. 2) dates from two periods in the history of Volga Bulgaria – pre-Mongol and Golden Horde (XII–XIV centuries). The hillfort was built in the XII century and inhabited until the XIV century (Gubaidullin, 2017). The settlement was first described by Vyacheslav who noted "... fortified settlements, burial mounds and other ancient earth mounds in Kazan governorate" (Vyacheslav, 1874). In 1877, Shpilevsky noted that the territory of the Lukovskoe settlement was plowed. In 1891 and 1909, Akhmarov examined the settlement and noted the plowing of the archaeological site, which allowed him to collect a rich material. He notes the destruction of the monument from the south and west under the influence of Kubnya river, "the area of the settlement decreases every year". Akhmarov determined the length of the shafts – 110 fathoms (about 240 m) (Akhmarov, 1894, 1910). In 1950, Kalinin investigated the settlement and reported the area of the monument was 23 000 m^2 and that fortifications had shortened by 40 m since the 1891 survey (Kalinin, 1954). In the 80s–90s of the 20th century, the fortifications of the Lukovskoe settlement were examined by archaeological expeditions of Iskander Lerunovich Izmailov and Airat Maratovich Gubaidullin. By comparison with the data of Nikolay Filippovich Kalinin expedition monument area significantly decreased due to continuous riverbank caving (Gubaidullin, 2017).

Thus, the results of field studies for almost 130 years indicate both the anthropogenic impact on the territory of the monument and its destruction due to the dynamics of the Kubnya river.

UAV surveys (2017, 2018), orthophotos (2017) and aerial photographs (1958) improved identification of changes in landscape. For 59 years, significant changes caused by riverbed displacement occurred in the study area (Fig. 3).

The site and the fortifications was previously destroyed as a result of channel deformations from the northern side, as evidenced by the presence of a crescent lake. Its convex side, on which the main energy of the stream falls, is directly adjacent to the territory of the settlement and forms a ledge. This is confirmed by literary sources, which describes the destruction of the settlement by fluvial processes and indicates the change of fortifications length – 240 m in 1891 (Akhmarov, 1894), in 1950–200 m (Kalinin, 1954). By the middle of the

Figure 2. Location of the Lukovskoe fortified settlement on the Kubnya river.

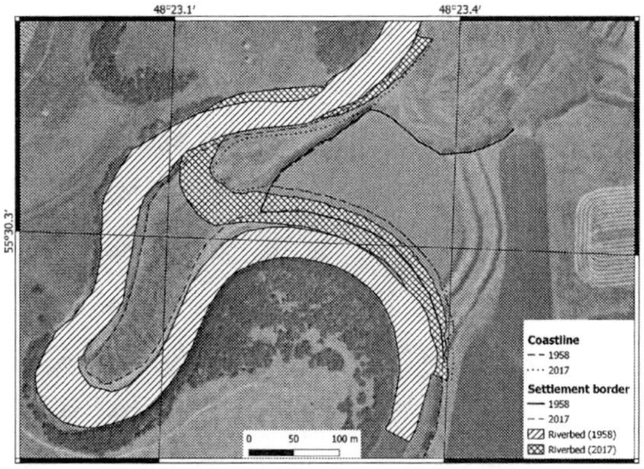

Figure 3. River channel change at the Lukovskoe fortified settlement (1958–2017).

20th century at least 30 % of the settlement area and 50 % of fortifications were lost. The original form of the settlement cannot be restored. The northwestern part of the monument is bordered by a low ledge (5–6 m). The ledge is covered by grass and no evidence of recent exogenous processes was observed. However, few paths created by cattle are cutting the ledge and leading to erosion.

The southern part of the settlement is influenced by the movement of the Kubnya river. The rates of horizontal deformations of small rivers channels flowing in loose sediments vary, typically from 0.1 to 2 $m\,yr^{-1}$. This is 2–3 times less than average rate and 15–20 times less than the extreme values of bank erosion on large rivers (Reference). Erosion of the lower stream bank which consists of sand of the channel facies is eroded by the stream causing the upper part of the floodplain facies to collapse. The meander bend continues to extend into the settlement. From 1958–2017, the river bank retreated by an average of 18.8 m (0.32 m yr^{-1}). In total, 5530.5 m^2 of material was removed by the river which constitute approximately 18 % of the settlement in 1958. At present, the length of fortifications is 180 m. In the middle of the 1990s, the riverbed "shifted" 254 m to the northeast. Therefore, displacement of the river Kubnya in the south-east direction (0.15 m yr^{-1}) can be observed in the northern part of the settlement, which may soon lead to the resumption of the bank caving processes by the channel flows at this area.

4 Conclusions

The present study of the effect of fluvial processes on the Lukovskoe fortified settlement location showed river bank erosion accounted for a loss of approximately 1/5 of the settlement. Recreation and agricultural practices (pasture) also influenced the historical settlement. More detailed study of the settlement is necessary to assess the risk of natural and anthropogenic impacts on the settlements to determine the implement protection zones for the monument and develop recommendations for security and rescue work.

Data availability. We cannot provide historical aerial photography data, as they contain State Secret. The data of the own survey with the UAV can be provided to all interested scientists upon request to the correspondence author.

Author contributions. The authors carried out work on field surveys with UAVs (PK, AK), UAV data processing and mapping (AG), inroduction and methods section and historical data analisys (IG), description of fortified settlement state and fluvial processes intensity analisys (BU).

Competing interests. The authors declare that they have no conflict of interest.

Special issue statement. This article is part of the special issue "Land use and climate change impacts on erosion and sediment transport". It is a result of the ICCE Symposium 2018 – Climate Change Impacts on Sediment Dynamics: Measurement, Modelling and Management, Moscow, Russia, 27–31 August 2018.

Acknowledgements. This work was supported by Russian Foundation for Basic Research (project No. 18-09-40114) (field survey, data processing and analisys) and performed according to the Russian Government Program of Competitive Growth of Kazan Federal University (providing historical information).

Financial support. This research has been supported by the Russian Foundation for Basic Research (grant no. 18-09-40114).

References

Akhmarov, G.: The Yapanchino fortified settlement on the river Kubna, Society of Archeology, History and Ethnography at Kazan University, XII, 179–182, 1894.

Akhmarov, G.: Report on a trip with an archaeological purpose in the summer of 1909 to Sviyazhsky and Tetyushsky districts of Kazan governorate, Society of Archeology, History and Ethnography at Kazan University, XXV, 179–182, 1910.

Fernández-Hernandez, J., González-Aguilera D., Rodríguez-Gonzálvez P., and Mancera-Taboada J.: Image-based modelling from unmanned aerial vehicle (UAV) photogrammetry: an effective, low-cost tool for archaeological applications, Archaeometry, 57, 128–145, https://doi.org/10.1111/arcm.12078, 2015.

Gafurov, A. M.: Small catchments DEM creation using Unmanned Aerial Vehicles, IOP Conf. Ser.-Earth and Env. Sci., 107, 12005, https://doi.org/10.1088/1755-1315/107/1/012005, 2018.

Gainullin, I. I., Khomyakov, P. V., Sitdikov, A. G., and Usmanov, B. M.: Qualitative assessment of the medieval fortifications condition with the use of remote sensing data (Republic of Tatarstan), Proceedings of SPIE – The International Society for Optical Engineering, 10444, 104440X, https://doi.org/10.1117/12.2279136, 2017.

Gubaidullin, A.: Fortification in the Middle Volga in the X – the first half of the XVI century (based on archaeological research), Doctoral thesis, 2017.

Ivanov, M. A., Prischepov, A. V., Golosov, V. N., Zalyaliev, R. R., Yefimov, K. V., Kondratieva, A. A., Kinyashova, A. D., and Ionova, Yu. K.: Method of croplands dynamics mapping in river basins of the European part of Russia for the period of 1985–2015, Sovremennye Problemy Distantsionnogo Zondirovaniya Zemli iz Kosmosa, 14, 161–171, https://doi.org/10.21046/2070-7401-2017-14-5-161-171, 2017.

Kalinin, N. F. and Khalikov, A. H.: Results of archaeological work for 1945–1952 years, Works of the Kazan branch of the USSR Academy of Sciences, Historical Sciences Series, Kazan, Tatar book publishing house, 126 pp., 1954.

Nicu, I. C.: Natural risk assessment and mitigation of cultural heritage sites in North-eastern Romania (Valea Oii river basin), Area, 51, 142–154, https://doi.org/10.1111/area.12433, 2018.

Usmanov, B., Yermolaev, O., and Gafurov, A.: Estimates of slope erosion intensity utilizing terrestrial laser scanning, Proc. IAHS, 367, 59–65, https://doi.org/10.5194/piahs-367-59-2015, 2015.

Usmanov, B., Nicu, I. C., and Gainullin, I.: Monitoring and assessing the destruction of archaeological sites from Kuibyshev reservoir coastline, Tatarstan Republic, Russian Federation. A case study, J. Coast. Conserv., 22, 417–429, https://doi.org/10.1007/s11852-017-0590-9, 2018.

Vyacheslav, N.: Notes on fortified settlements, burial mounds and other ancient earth mounds in Kazan governorate, Kazan governorate printing house, 33 pp., 1874.

Yermolaev, O. P. and Avvakumova, A. O.: Cartographic-geoinformational estimation of spatio-temporal erosion dynamics of arable soils in forest-steppe landscapes of the Russian Plain, Erosion and Sediment Yields in The Changing Environment, IAHS Publication, 356, 332–337, 2012.

Yermolaev, O. P., Gafurov, A. M., and Usmanov, B. M.: Evaluation of Erosion Intensity and Dynamics Using Terrestrial Laser Scanning, Eurasian Soil Sci., 51, 814–826, https://doi.org/10.1134/s1064229318070037, 2018.

Proc. IAHS, 381, 37–47, 2019
https://doi.org/10.5194/piahs-381-37-2019

Magnitude and frequency of debris and slush flows in the Khibiny mountain valleys, Kola Peninsula, NW Russia

Ekaterina V. Garankina[1], **Vladimir R. Belyaev**[2], **Fedor A. Romanenko**[1], **Maxim M. Ivanov**[2], **Natalia V. Kuzmenkova**[3], **Artem L. Gurinov**[1], **and Egor D. Tulyakov**[1]

[1]Department of Geomorphology and Paleogeography, Faculty of Geography, Lomonosov Moscow State University, Moscow, 119991, Russia
[2]Laboratory for Soil Erosion and Fluvial Processes, Faculty of Geography, Lomonosov Moscow State University, Moscow, 119991, Russia
[3]Department of Radiochemistry, Faculty of Chemistry, Lomonosov Moscow State University, Moscow, 119991, Russia

Correspondence: Ekaterina V. Garankina (evgarankina@gmail.com)

Published: 1 August 2019

Abstract. Available results for five studied valleys of the Khibiny Mountains, Kola Peninsula, suggest that slush flows and, possibly for some valleys, typical debris flows with lower frequency, are a leading mechanism for downstream sediment delivery and valley floor topographical formation. Typical fluvial topography in slush flow-affected basins is extremely suppressed or nonexistent, since under such conditions, stream channels are unable to rework slush flow deposits. The recovery phase of fluvial topography can serve as an indicator of the magnitude and time passed since the last extreme event. A combination of grain size analysis, radionuclide fingerprinting with the ^{232}Th content in the finer-grained sediment matrix (size < 10 mm) and ^{14}C dating, were applied to reveal the age and common structure of debris and slush flow environments and to investigate the main factors in their lithodynamics. Those helped to estimate transportation distances and capacities of the flows and the amount of fluvial reworking of its deposits with time. Application of radiocarbon dating to determine absolute ages (about 30 dates) of stabilization periods for the colluvial cones, mountain fans and valley bottoms and integration with other available chronological data provided a basis for distinguishing several stages of decreased activity of debris and slush flows and extreme slope failures through the second half of the Holocene. Field mapping and remote sensing data interpretation revealed spatial distribution patterns of debris and slush flows. Geomorphic analysis of large relic landforms in valley bottoms confirms, in general, the case for a significant reduction of debris flow magnitude since the last deglaciation and distinct shift to slush flow processes with much lower clastic content. A reliable chronology of the early events is yet to be obtained representing a challenging problem for future research.

1 Introduction

The largest mountain massif of the Kola Peninsula – Khibiny – is comprised of Devonian alkaline intrusions (Pozhilenko et al., 2002). Compact plateau-shaped mountains (up to 1201 m a.s.l.), with relatively steep slopes, are dissected by numerous deep (100–400 m) erosional valleys, glacial troughs and cirques, and tectonic lineaments. Extremely rich in minerals, especially in apatite and nepheline ores and rare earth elements (Fig. 1), Khibiny have been intensely explored since the 1920s. Recently, the area has become a focus of a rapidly growing recreational industry. At the same time as being an arena of widespread natural hazards, its industrial and civil infrastructure (ski resorts, newly established National park, quarries, mines, plants, roads, etc.) is strongly affected by snow avalanches, rockfalls, screes, and debris flows (Belyaev et al., 2015). The most frequently observed types of the latter are snowmelt period slush flows and rainfall-induced medium magnitude debris flows (Perov et al., 2017).

Slush flow represents a specific gravitational flow of a water-saturated mixture of snow and a limited amount of clastic sediment (up to 12 %, clasts size up to 1–2 m) occurring in low-order stream channels (Perov, 1996; Bozhinsky and Myagkov, 2001; Fleishman, 1978). Slush flows are classified as subtypes of wet snow avalanches or debris flows or as independent phenomena between the latter two (Hestnes, 1998; Eckerstorfer and Christiansen, 2012).

Resembling sets of hazardous episodes, some with fatal consequences, are reported for other mountainous Arctic and Subarctic environments in Russia, Scandinavia (Nyberg, 1989; Hestnes and Kristensen, 2010; Jonsson and Gauer, 2014) and Northern America (André, 1995; Larocque et al., 2001; Relf et al., 2015, etc.). In Japan, more than 90 casualties due to slush flows have been registered since 1945 (Kobayashi et al., 1994). To estimate potential risks of these hazards, a thorough understanding of their nature and dynamics is required.

The Khibiny Mountains provide a unique dataset of more than 200 catchments affected by slush flows for the last centennial (Bozhinsky and Myagkov, 2001). Fifty years of survey mainly concentrated on monitoring the observed events and developing protection recommendations for mining infrastructure (Chernous, 2006) whilst the distribution, magnitude and frequency of such hazardous events in the past remained beyond the scope of exploration. However, large-scale bottom and piedmont fans found in the majority of small valleys do not always correspond to a glacial origin suggesting previous periods of much higher debris flow intensity. Given the lack of spatial data and absolute age determinations, improved understanding of correlations between such deposits and landforms is essential for reliable paleogeographic reconstructions. In this article, we therefore try to evaluate dynamics of debris and slush flows during the Holocene in the Khibiny mountain valleys.

2 Methods

Comprehensive field investigations including detailed descriptions of several geological sections were accomplished in five valleys of the Kuna River, Malaya Belaya River, and Bolshaya Belaya River basins (Fig. 1). Geomorphic interpretation of high-resolution satellite imagery from public services, aerial photography both archived and obtained by UAV DJI Phantom III, and 1 : 50 000 scale topographic maps assembled evidence for relic and active debris flows. The relative ages of these phenomena were estimated based on both morphological indications and vegetative cover (Rudinskaya et al., 2018).

Grain size analysis, radionuclide fingerprinting, and ^{14}C dating were applied to reveal the age and common structure of the debris flow environments, and the main factors affecting their lithodynamics. Fresh slush flow accumulations were sampled for assessment of deposition and fluvial reworking rates. Surface sampling of channel beds and banks, valley bottoms and sides helped to specify the modern patterns and agents of basin-scale sediment redistribution based on ^{232}Th content. Cross-sections of superimposed debris and slush flow, fluvial and colluvial, lacustrine and peat deposits in terraces and fans between 1 and 10 m above the streams were sampled to establish both distinctive features of each genetic type and their age.

We applied a standard procedure for the finer-grained sediment matrix (size < 10 mm) dry sieving with the Fritsch Analysette 3 PRO shaker to separate fractions of < 0.05, 0.05–0.1, 0.1–0.25, 0.25–0.5, 0.5–1, 1–2, 2–5, and 5–10 mm. For gamma-spectrometric analysis, each fraction was ground and weighed into 1 g subsamples. Gamma activities of ^{232}Th daughter radionuclides ^{212}Pb and ^{228}Ac were measured with the ORTEC GEM-C5060P4-B HPGe detector. Alpha activities of $^{228,232}Th$ isotopes of sequentially extracted aliquots of the most active samples were measured with the ORTEC Alpha-Ensemble-2 with detector ENS-U900 type UL-TRA-AS displaying the content and ratio of ^{212}Pb and ^{228}Ac and corresponding physicochemical forms of ^{232}Th (Ivanov et al., 2019).

A limited geochronological framework was established by ^{14}C dating of buried humic and peat layers between superimposed debris flow deposits. The results were compared to the published Holocene chronology of local slope processes. Scintillation radiocarbon dating was performed at the Laboratory of Radiocarbon Dating and Electron Microscopy, IG RAS, Russia.

3 Results

In 2015, we started investigations of the frequency and age of debris and slush flow events in the Khibiny (Garankina et al., 2018) and Lovozerskiye Tundry (Rudinskaya et al., 2018). Field observations revealed both modern and older corresponding landforms: valley floor and stream bank incisions, ridges, lobes and hummocks, debris covers, terraces, and fans. The surfaces of relatively fresh landforms have no or poor vegetation cover (fragmentary mosses, lichens, herbs, shrubs) and do not exceed 1 to 2 m in height and 0.02 km^2 in area. Older features are larger (up to 10 m high and 0.05–0.4 km^2 in area) and are mostly well forested or covered by thick tundra. Deposits vary from large boulders and blocks to sandy silts depending on the geomorphic position above the streambed, the distance from the sources, age, and flow energy. Usually, up to three levels of debris or slush flow accumulations of different age can be distinguished. Along the middle reaches of 2nd and 3rd order streams, they are traced as sequences of inclined terraces. In the lower reaches, they appear as a series of humic or peat layers buried between coarse deposits in slush flow fans.

Most fresh slush flow deposits were found days to months after their accumulation in valleys of the Alyavumjok, North-

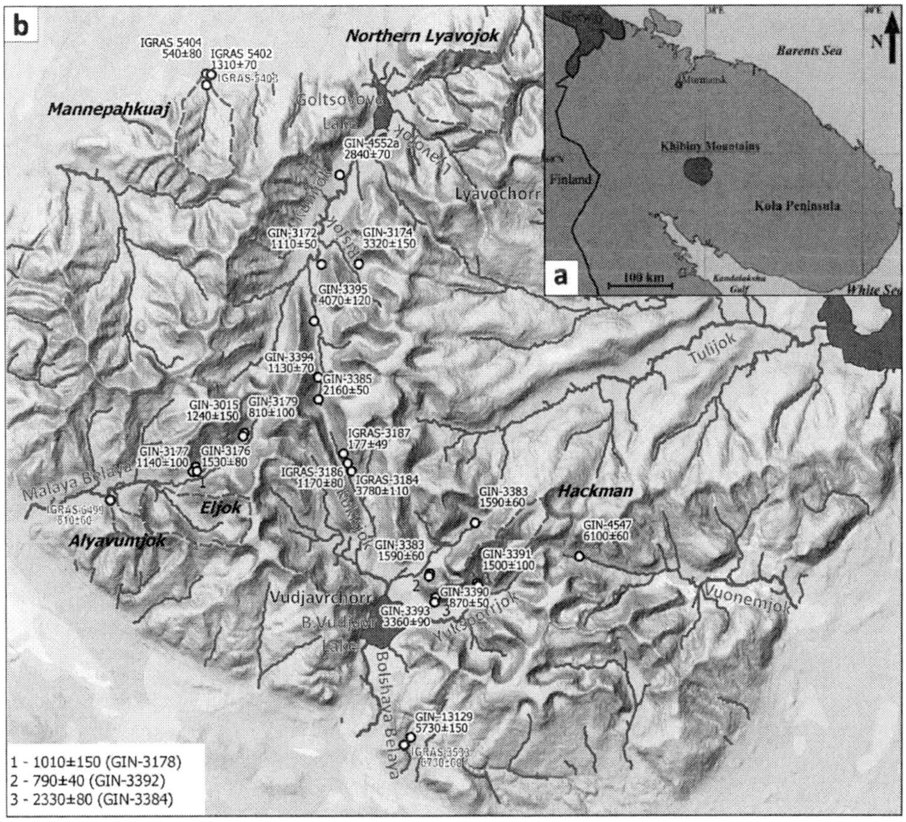

Figure 1. Location of the Khibiny Mountains on the Kola Peninsula, NW Russia **(a)** and case study basins **(b)** (source: © Google global topography, infrastructure and settlements). Panel **(b)** shows the distribution of a dataset of ^{14}C dates of buried organic-rich horizons in colluvial and debris flow fans (after Sulerzhitskiy et al., 1986; Vashchalova, 1987; Bozhinsky and Myagkov, 2001; Vladychensky et al., 2007; Kosareva, 2007; Romanenko et al., 2011, with alterations).

ern Lyavojok, and Mannepahkuaj rivers. In the latter, they formed thin debris covers, individual hummocks up to 1.5 m high and ridges 7–8 m long during the 2016 early spring snowmelt (fresh silty gravel and boulder loads at the confluence of main headwaters were registered at the end of June). Analysis of freely available remote sensing data allowed determination of the associated slush flow discharge periods in the spring of 2014 and at least once before, in 2012–2013. A slush flow from the main Northern Lyavojok tributary constituted a larger debris fan protruding into the forested valley bottom. Based on the sparse vegetative cover, this slush flow occurred less than 10–15 years ago.

At the footslope of the Vudjavrchorr Mountain, Southern Khibiny, a deep V-shaped erosional incision cuts into the 150 m thick glacial drifts. Here, the fan has several prominent age generations represented by a series of inclined lobes with ~ 2 m difference in elevation between each other (Fig. 2). Their surfaces are armored by sand and gravel with some boulders with small clastic ridges atop. The lobe nearest to the channel is devoid of vegetation, while the most remote one is already covered by dwarf willows and birches. This can be considered as evidence of repeated slush flow activity over the last centuries. The youngest non-vegetated fan surface is morphologically fresh and is observed on 1958 airborne photos. It can be speculated that it was formed over the last 100 years (Table 1). In close vicinity, there is another similar erosional cut but under the dense spruce forest, suggesting it is of earlier origin. Corresponding older fan is also completely vegetated and consists of deposits of two ages.

Evidence of relatively young slush flows of substantial volume is observed in the Malaya Belaya River basin. Two small catchments (Alyavumjok and Eljok) on its left side have large unvegetated fans. The valley floor of the former is almost devoid of loose deposits, and the latter has a deep (up to 35 m) V-shaped incision in the middle reaches. Debris fans 200–250 m in radius with boulder paving are shown on the first topographic maps from the 1930s. Thus, they are at least 90 years old, being superimposed on forested and even more

Figure 2. Several generations of debris flow fans distinguished by relative elevation and vegetation succession stages below the V-shaped cut into moraine under the southern slope of the Vudjavrchorr Mountain.

Table 1. Vegetation characteristics for indicating the age of debris or slush flow deposit, established for the forest belt in the Khibiny Mountains (after Bozhinsky and Myagkov, 2001).

Age, years ago	Landscape characteristic
< 10	Sparse grasses and dwarf shrubs
10–20	Dense grasses and dwarf shrubs with some young tree shoots
20–50	Dominance of crooked birch forest
50–100	Mixed forest dominated by deciduous trees
> 100	Mature forest dominated by spruce

expansive older fans, which displays repetitive activity of extreme slush flows. The older Alyavumjok fan overlapped the Malaya Belaya trough floor and forced the main river channel to shift its position significantly and, thus, a meander up to 350 m in radius to form. The frequency of such disastrous events can be estimated by ^{14}C dating of buried soils within the forested piedmont areas. A well-developed humic horizon under the post-slush flow flood sediment shows that the last extreme episode affected this part of the Alyavumjok fan less than 800 years ago (Fig. 5a) – subsequent to the occurrence of the above-mentioned meander.

The distribution of slush flow patterns and landforms is strongly controlled by tectonic structure. For example, the feather-shaped drainage network of the Mannepahkuaj Basin induces tributary slush flow runout zones in the main valley bottom. Such coarse debris dams occasionally block the main channel forcing the stream to filtrate through and leading to further outbursts. Downstream, such "wavy" structure of deposition repeatedly induced normal debris flows with unequal transportation distances of the sediment load. This caused a series of debris fan lobes of different sizes to form at the forested piedmont (up to 6 km from the sources). Ages of embryonic paleosols (Fig. 5a) in two successive deposition zones, 0.5 km apart, reflect high-magnitude debris flows at least twice per millennia and probably much more often (Garankina et al., 2018).

It has been determined that slush flow fans or other landforms in their deposition zones under certain conditions become an important control of the evolution of fluvial topography. In the higher order river basins, it is often challenging to recognize zones of dominant slush flow origin, transport or deposition. Smaller and less complex catchments (mostly of the 1st order streams) present distinctly confined slush flow morphodynamic zones. For more detailed investigations of the slush flow–fluvial interactions, several basins were chosen for grain size and radionuclide content evaluation.

In fresh slush flow and alluvial deposits at the Mannepahkuaj streambed (Fig. 3), sandy and silty fractions gradually increase downstream. In the lower reaches, inhomogeneity is caused by a combination of variable transportation capacities of different magnitude debris and slush flows and the amount of fluvial reworking of their deposits. Older fans and terraces in the middle reaches show much coarser composition, probably, due to the higher transportation capacity of slush flows in the past. This is in agreement with the much larger extent of correlated landforms (up to 10 m relative height). At the relic piedmont fan, the sand content increases, however, the grain size composition is generally finer than in modern slush flow deposits. This may be explained by the

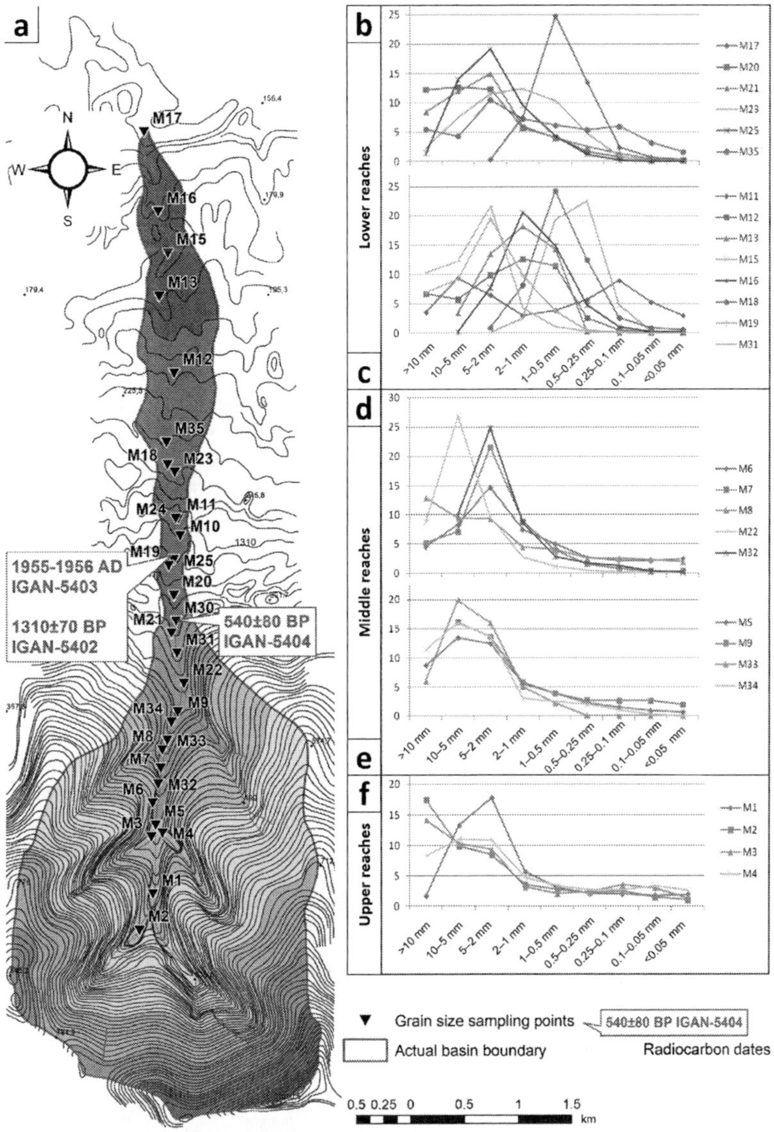

Figure 3. Grain size data of slush flow deposits in the Mannepahkuaj Basin: **(a)** sampling points on the elevation map. Distribution of granulometric fractions in samples from relic valley bottom features **(b, d)** and fresh accumulations in the river channel **(c, e, f)**.

greater area of the relic fan (approximately 8 km^2, equivalent to the entire mountainous area of the basin) compared with the narrow strip of modern accumulation along the channel. Thus, stream bifurcation and rapid reduction in transportation capacity with distance from the mountains is likely to have occurred in the past. Hence, the main tendencies in longitudinal differentiation are gradual channelization of slush flow deposition, leaving large relic landforms inactive and the resultant shift in clastic material sorting, which substantially decreases with time.

Recent publications report the use of ^{232}Th radionuclides as tracers for distinguishing the main exogenous processes of sediment transport in the Khibiny valleys. For the Hackman basin with a strong ^{232}Th signal caused by radioactive mine tailings, gamma spectrometry analysis revealed definite ^{232}Th concentrations in valley bottom deposits in comparison with adjacent colluvial slopes and channel bed alluvium.

This provides a basis for confirming episodic powerful removal and mixing of material along the valley by slush flows (Ivanov et al., 2018). The contribution of constant water flow within the stream channel is limited to washing the fine fractions out. The latter, according to the alpha spectrometry (Ivanov et al., 2019), make the most relevant contribution to the total radioactivity of the samples and can cause positive radioactive anomalies in sedimentation basins (deltas, lakes, artificial ponds) outside the catchments. Thus, secondary redeposition of the ^{232}Th-rich fine fractions is possible downstream, along the receiving Yuksporrjok River and towards the Bolshoj Vudjavr Lake.

The Northern Lyavojok Valley – with confirmed but scattered ^{232}Th bedrock sources – was chosen for fingerprinting due to its potential for tracking the transport of radioactive sediment down to the Goltsovoye Lake with the frequent influence of slush flows in the background. The last (late Weichselian) continental ice cover left the Khibiny northern valleys almost devoid of glacial drifts; however, widespread thick debris fans, ridges, and elevated fragments of former valley floors prove high depositional intensity here in the past, interrupted by several major incisions, probably due to the subsidence of the local base level during deglaciation and continuing tectonic uplift (Garankina et al., 2018). In addition to the surface sampling of the valley, a bottom core obtained from the northern nearshore part of the Goltsovoye Lake containing a sediment record of at least the last 7 kyr was examined for radionuclide content. Elevated ^{232}Th values in the lacustrine sediments are believed to identify periods of intensified delivery of ^{232}Th-rich alluvium from the valley. Granulometric differentiation of sediments along the transportation path probably also takes place and can influence radioisotope concentrations (Fig. 4a). Gamma-spectrometry measurements of unsupported ^{210}Pb ($^{210}Pb_{ex}$) concentration produced a complex depth distribution curve designating variable input of the radionuclides into the lacustrine environment. This depth profile revealed that the upper 0.3 m of sediments accumulated in the lake during the last 200 years (Fig. 4b). The high variability of ^{232}Th activity detected in the lacustrine deposits may well be explained by periodic increased input of sediment into the lake caused by slush flows or extreme floods in the main eastern tributaries – Lyavojok and Northern Lyavojok.

4 Discussion

Magnitude and frequency analysis and dating of key activity periods for slope processes (screes, rockfalls), snow avalanches, debris and slush flows represent important areas of research for improving understanding of morphodynamics in the Khibiny. Work has already been undertaken to date phases of decreased activity of catastrophic processes (Perov, 1971; Vashchalova, 1987; Bozhinsky and Myagkov, 2001; Vladychensky et al., 2007; Kosareva, 2007; Romanenko et al., 2011; Nikolaeva, 2014; Shilova and Romanenko, 2016; Nikolaeva et al., 2016). These studies employed dendrochronology, airborne photograph interpretation, and landform analysis. Unfortunately, all these approaches are limited in terms of both temporal resolution and maximum possible dated age.

Sets of geobotanic landscape indicators for slush flow fan and colluvial (rockfall or scree) cone ages were distinguished (Table 1). In general, the longer the interval without catastrophic geomorphologic events, the denser the vegetation cover on a fan. However, it must be borne in mind that such indicators are inapplicable for the tundra landscapes occupying a large part of the mountain slopes and summits with slush flow-affected catchments.

Perov (1966, 1971) first used a dendrochronological approach in the Kunijok River basin, and this was later improved by Olga I. Budarina and Galina G. Sapunova (Bozhinsky and Myagkov, 2001). This approach is seriously limited by dependency on local hydrothermal conditions, soil, and substrate, poorly readable tree rings, and dramatic loss of information with increasing age. The most reliable dates can be obtained by measuring young shoots from the bases of trees fallen or deformed by flows. On this basis, massive activations of slush flows were established for 1943, 1946, 1950–1952, 1960, 1966, 1969, 1977, and 1987 (Perov, 1966; Sapunov, 1991; Bozhinsky and Myagkov, 2001). Ananiev (1998) reported massive slush flow traces in the Hackman Valley that had almost damaged the infrastructure of the Rasvumchorr mines in May 1995.

Avalanches in the Kola Mountains typically occur annually – snow bodies are regularly found in valleys and can remain unthawed till the following winter. Rockfall, scree, and debris flow processes normally take place much less frequently. However, huge block fields – seismic rockfall bodies – are widespread on the foothills. In several basins of the Northern and Western Khibiny, dramatic imprints of extreme events represented by substantial clastic volume but which was transported for relatively short distances (0.5–2 km) are regarded as traces of seismically triggered catastrophic events. Combinations of poorly sorted large clastic bodies (block sizes > 1–3 m) occupying cirque and valley bottoms and thick debris fans blocking adjacent small tributary valleys or thrown into the main troughs provide evidence for simultaneous or consequent rockfalls and debris flows (Garankina and Lukashov, 2018). Previously, Romanenko et al. (2011) reported a bog dated about 5.7 ka that was formed due to damming of the small valley by a debris flow fan in the southern Khibiny. A seismic event had most likely triggered a rockfall causing catastrophic water outburst from the lake in the headwaters. Slope mass movements, in contrast, cover most parts of the mountains, but result in much slower and yet distinct translocations of colluvial mantle and soils burial. Those features are revealed by mobile plant communities and the dynamics of cryogenic microtopography (Garankina, 2013).

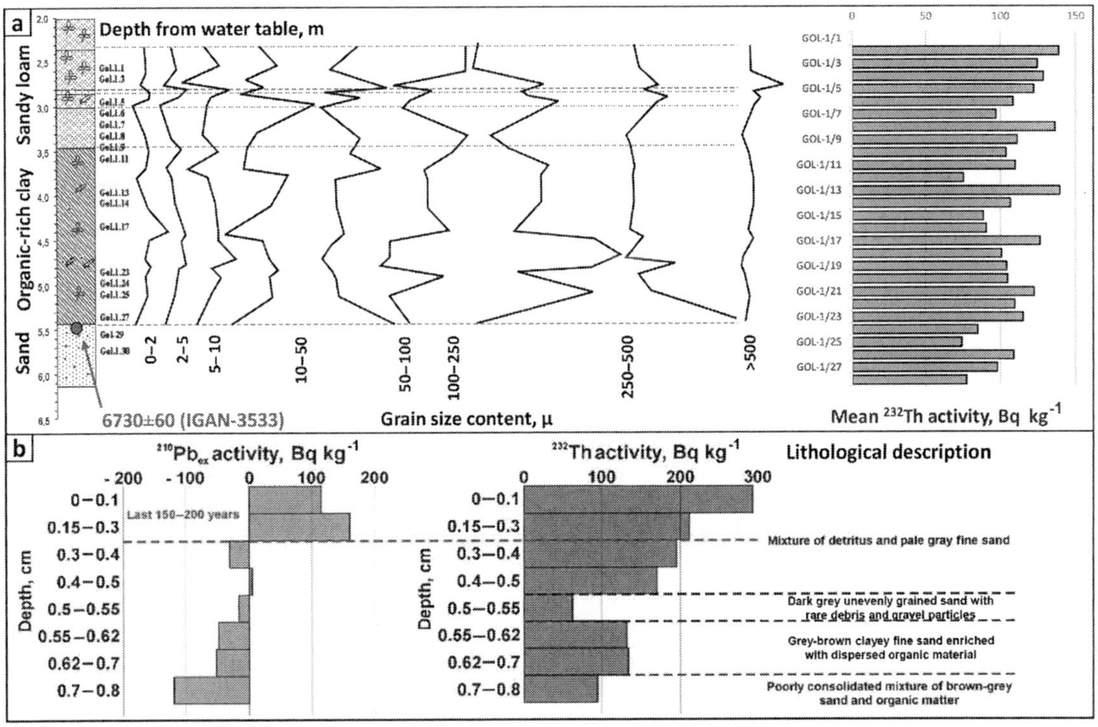

Figure 4. Redistribution of ^{232}Th radionuclides in lacustrine sediments of the Goltsovoye Lake **(a)**. Depth distribution of $^{210}Pb_{ex}$ and ^{232}Th in the upper part of the profile **(b)**.

The introduction of the radiocarbon method became a turning point in progressing understanding of the age of all of the above-mentioned processes. However, dating of the catastrophic events themselves is still rather difficult. Their age limits are most reliably obtained by testing organic compounds formed during periods of relative stability, which separate the phases of extreme activities. Such materials include peat and primitive soil layers or even lenses of mineral substrate with little content of humified matter. Such layers form on stable surfaces and later become buried by younger clastic deposits. For the events themselves, such dates determine lower (prior to the event) time marks or, alternatively, both lower (prior) and upper (post-event) marks in sections with several datable organic horizons separating several clastic layers.

Vashchalova (1987) assembled the first series of ^{14}C dates within the interval from 810 to 6100 ^{14}C years (Fig. 1) in the Khibiny Mountains and successfully determined periods of decreased snow avalanches, screes, and rockfalls dynamics. Combined with the dates acquired during this particular study (Fig. 5a) and previous research (Romanenko et al., 2011), a dataset of up to 30 ^{14}C dates was compiled (Fig. 5b). This dataset suggests that in the second half of the Holocene, the intensity of catastrophic slope and slush flow processes could have increased during the periods BCE 3600–2800, 1500–1250, 850–700, BCE 150–CE 300, and since CE 1350 years. The initiation of the potential latest activity seems to coincide with the beginning of the so-called Little Ice Age of the XIV–XIX centuries. In the case of the dynamics of snow avalanches, this pattern is explained by the better snow supply (Vashchalova, 1987). Intensification of slope processes and debris flows can also be proposed for the beginning of the Holocene climatic optimum between 7.5 and 6 ka (BCE 5500–4000 years). However, there is a certain spatial inhomogeneity of stabilization periods to be found over the Khibiny area. For the historic times, definite episodes of intensification of catastrophic processes can be determined for individual basins not responding to the regional signal and potentially caused by varying local geomorphic and snow cover conditions.

The current insufficient number and spatial coverage of the available dates point to the need for collecting more data on debris flow phenomena over the entire Late Glacial–Early Holocene. Nevertheless, the extensive occurrence of distinctive large relic landforms and thick bottom deposits without any detectable organic matter indicates a substantially higher magnitude of debris flow activity in the distant past. Most likely, these debris flows functioned in colder and snowier a

№	Basin	Section	Depth, cm	Material	Lab number	^{14}C, yrs BP (±1σ)
1	Mannepahkuaj River, lower reaches	2 m slushflow terrace	30–40	TOC	IGAN-5403	101,40 % ±2,23 %
2			60–70	TOC	IGAN-5402	1310±70
3		2 m slushflow terrace	100–110	TOC	IGAN-5404	540±80
4	Alyavumjok River	Relic debris flow fan	35–40	Peat	IGAN-6499	810±60
5	Goltsovoye Lake	Bottom sediment core	540–550	TOC	IGAN-3533	6730±60

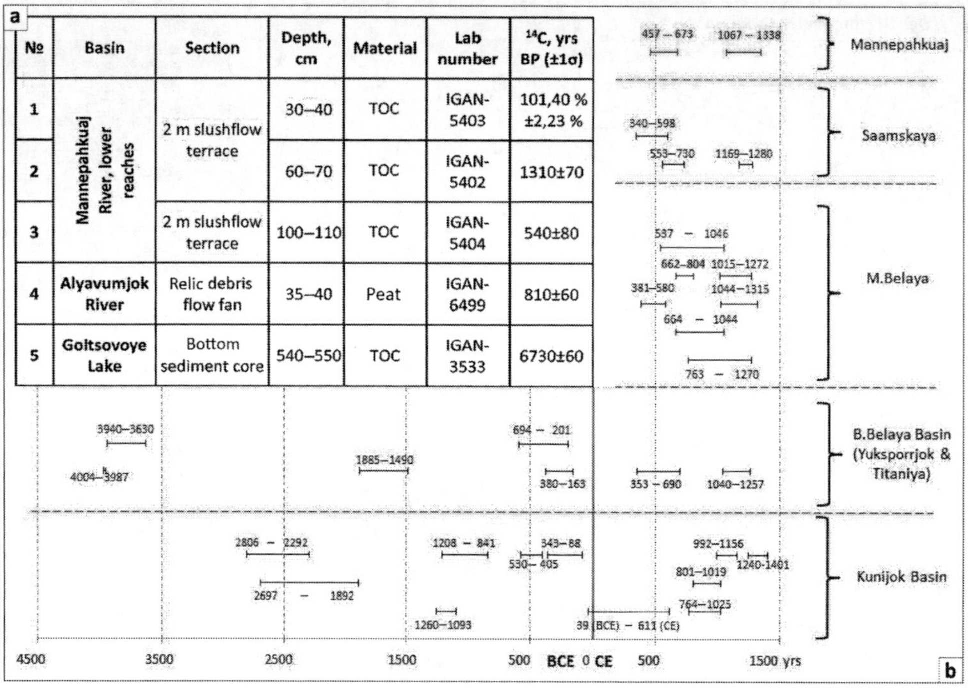

Figure 5. Holocene geochronology for the valleys in the Khibiny Mountains. New ^{14}C dating results **(a)**. Intervals (calibrated ages) of buried soil horizons in debris flow fans and slope cones **(b)**.

environments during the last deglaciation stages. Here, it is probable that the remaining mountain glaciation in the affected valley headwaters caused periodical moraine-dammed lake outbursts and associated glaciogenic debris flows.

Interpretation of the evidence can be substantially improved by comparing data from other natural paleoarchives reflecting the general tendencies of landscape evolution. Traces of catastrophic events fixed in lake bottom sediment records can be revealed by supplementary approaches. For instance, recent work by a research group from the Kola Scientific Center RAS (Nikolaeva et al., 2016) revealed a catastrophic outburst into the Imandra Lake from the surrounding mountainous catchments as indicated by a coarse clastic breccia-type layer within the gyttja deposits. The authors associate this outburst with the seismic event between 6.5 and 5.6 ^{14}C ka. Nikolaeva (2014) also attribute folding and faulting deformations in varved clays of the Vuonemjok valley to a seismic event dated to Allereod; however, opposing interpretation of such disturbed lacustrine lamination as a result of a glacial outburst flood or debris flow impact was also proposed by Gorbatov and Kolesnikov (2017).

A prominent 6 cm thick layer of clear grey silt was found within organic deposits southward of the Goltsovoye Lake. As reported by Shilova and Romanenko (2016), this layer is younger than 5.5 ka and contains a substantially lower amount and taxonomic variability of diatoms, although without significant differentiation of their species. In addition, higher amounts of terrigenous material are characteristic for the underlying Early Holocene sandy facies. All of the above can be taken as evidence of some catastrophic (probably, debris flow) events within the basin, which caused fast and large-scale inputs of mineral sediment into the lake mainly at the beginning and occasionally in the second part of the Holocene.

The erosional potential of slush and debris flows can be estimated based on geomorphic footprints. Deep V-shaped incisions widespread in the upper reaches of the 1st order streams, which are almost dry during most of the warm season, attest to an origin resulting from an agent other than fluvial. In places, these features incise even bedrock, revealing the substantial intensities of the eroding flows. Normally, the catchment areas above them (cirques, niches, etc.) provide abundant snow and water supply and serve as modern slush flow sources. Downstream in the major valleys, several incisive cycles are imprinted in the bottom topography. Based on the relatively low discharge rates and no geomorphic indication of active erosion by fluvial processes, those features should also be associated with slush flow episodes, probably caused by climatic oscillations, or sometimes, local base level decrease.

A trend of reduction in the magnitude of debris flow phenomena since the last deglaciation is observed. During the early stage of the Holocene widespread debris flow deposition was dominant, leaving large geomorphic footprints inconsistent with modern runoff conditions. Most likely the largest of those were caused by proglacial lake outbursts. Later, debris flow transportation capacity and frequency decreased markedly, probably due to the reduction of both water and sediment sources. As a result, debris flows were largely replaced by slush flows with much lower clastic content, which resulted in much smaller modern accumulations. However, the higher erosional potential of slush flows has been causing ongoing incision into relic glacial and debris flow landforms.

5 Conclusions

Available results for the Khibiny Mountains suggest slush flows and typical debris flows with lower frequency are both leading mechanisms and underestimated agents of downstream sediment delivery and valley floor topographical formation during the Holocene.

1. In the 1st order streams, fluvial topography is extremely suppressed or nonexistent as stream channels are unable to rework slush flow deposits and are forced to adjust passively. In typical erosional valleys of the 2nd order, fluvial processes are also almost completely paralyzed by even minor deposition as a result of high-frequency slush flows. Here, streams are limited to reworking the finer fractions of slush flow fans, forming secondary alluvial features downstream. Small valleys of the 3rd and higher stream orders with glacial topography, are usually devoid of slush flows except for low-magnitude events of the wet avalanche type in riverbeds. Only rare major slush flow ejections from tributaries producing large superimposed fans in the main valley floor can influence their fluvial cycles. Such rare catastrophic events can lead to major river channel shifts and deep fresh-looking incisions.

2. The recovery phase of normal fluvial topography can be used to indicate the magnitude and time of the last extreme slush flow events by assessing the amount of fluvial reworking of associated deposits, transportation distances and capacities using grain size and radionuclide analyses.

3. The recurrence interval of medium magnitude slush flows does not exceed 10–30 years, which is in agreement with previously published monitoring data. Low magnitude events are even more frequent (up to every 1–2 years) in headwaters and tributaries of certain valleys. The frequency of extreme events is much lower and can be estimated as at least twice per millennia, according to radiocarbon dating, and is probably even higher.

4. Application of the radiocarbon method to determine the ages of stabilization periods for the Khibiny Mountains fans and valley bottoms provided a basis for distinguishing stages of non-uniform activity of slush flow phenomena during the second half of the Holocene, most likely controlled by climatic oscillations and, at places, by the fluctuations of local base levels. The intensity of catastrophic slope and slush flow processes increased during BCE 3600–2800, 1500–1250, 850–700, BCE 150–CE 300, and since CE 1350, and an even higher frequency of associated events can be speculated for the first half of the Holocene climatic optimum (7.5–6 ka) though direct geochronological evidence for this is still scarce.

5. Large geomorphic footprints inconsistent with modern runoff conditions and lacking any buried visible organic matter, in general, attest to a leading role of debris flows in sediment transport during the late deglaciation and the Early Holocene associated with repetitive breaches of moraine-dammed lakes and subsequent powerful glaciogenic debris flows. Later, a distinct decrease of debris flow transportation capacity and frequency occurred due to the depletion of both water and sediment sources. Here, debris flows were largely substituted by slush flows with much lower clastic content, which explains the much smaller modern sediment accumulations. However, the higher erosional potential of those phenomena caused a direct incision of glacial and debris flow landforms marked by a series of widespread deep superimposed cuts in valley bottoms. A transition period from the dominance of typical debris flows to that of the slush flows is yet to be determined more precisely by means of absolute geochronology and this represents the most challenging problem for future research in the area.

Data availability. The majority of the data presented here are from an ongoing and yet incomplete project. All datasets will be made (and partially already is) available on completion of the project and the results will be published with a link probably hosted by Russian Foundation for Basic Research. Please contact Ekaterina V. Garankina for further information.

Author contributions. All co-authors took part in field working and sampling. MMI and NVK performed radionuclide content studies. EDT carried out grain size analysis. FAR summarized previously accomplished geochronological data. ALG, VRB, and EVG designed cartographic and artwork and the latter prepared the manuscript with contributions from all co-authors.

Competing interests. The authors declare that they have no conflict of interest.

Special issue statement. This article is part of the special issue "Land use and climate change impacts on erosion and sediment transport". It is a result of the ICCE Symposium 2018 – Climate Change Impacts on Sediment Dynamics: Measurement, Modelling and Management, Moscow, Russia, 27–31 August 2018.

Acknowledgements. The authors are grateful to Andrey A. Lukashov, Olga S. Shilova and numerous MSU students for fieldwork assistance and data processing during the 2008 winter expedition and the 2015–2018 summer training courses of the Geomorphology and Paleogeography Dept.

Financial support. This research has been supported by the Russian Foundation for Basic Research project no. 17-05-00630 in part of fieldwork organization and radiocarbon dating and by the Geomorphology and Paleogeography Dept., Faculty of Geography, MSU (Governmental Assignment of no. AAAA-A16-116328100089-5) for assisting with grain size and radioisotope measurements.

References

Ananiev, G. S.: Catastrophic processes of land formation, MSU Publ, Moscow, Russia, 1998 (in Russian).

André, M. F.: Holocene Climate Fluctuations and Geomorphic Impact of Extreme Events in Svalbard, Geogr. Ann. A, 77, 241–250, 1995.

Belyaev, Y. R., Bredikhin, A. V., and Lukashov A. A.: Environmental geomorphologic safety of planning technical roads in the Khibiny Mountains, Georisk, 2, 50–57, 2015 (in Russian).

Bozhinsky, A. N. and Myagkov, S. M. (Eds.): Slushflows in Khibiny Mountains, MSU Publ., Moscow, Russia, 2001 (in Russian).

Chernous, P. A.: Destroying avalanche-like processes in the Khibiny Mountains: damage mitigation, in: Proceedings of the INTERPRAEVENT International Symposium Disaster Mitigation of Debris Flows, Slope Failures and Landslides, 25–27 September 2006, Niigata, Japan, 255–266, 2006.

Eckerstorfer, M. and Christiansen, H. H.: Meteorology, topography and snowpack conditions causing two extreme mid-winter slush and wet slab avalanche periods in high Arctic maritime Svalbard, Permafrost Periglac., 23, 15–25, 2012.

Fleishman, S. M.: Debris flows, Hydrometeoizdat, Leningrad, Russia, 1978 (in Russian).

Garankina, E. V.: Evolution of the cryogenic microrelief of the Subarctic low mountains, Earth's Cryosphere, 17, 3–16, 2013 (in Russian).

Garankina, E. V. and Lukashov, A. A.: Potential triggering role of seismic episodes for debris flow phenomena in the Khibiny (Kola Peninsula, Russia), Georisk, 12, 48–57, 2018 (in Russian).

Garankina, E. V., Belyaev, V. R., Belyaev, Y. R., Bondar, A. I., Gurinov, A. L., Ivanov, M. M., Romanenko, F. A., Rudinskaya, A. I., Shishkina, Y. V., and Tulyakov, E. D.: Slushflows as a land sculpturing agent in mountain valleys of the Kola Peninsula, Northwestern Russia, in: Practical Geography and XXI Century Challenges. IGU Thematic Conference, 4–6 June 2018, Moscow, Russia, Conf. Book. Part 1, 52–59, 2018.

Gorbatov, E. S. and Kolesnikov, S. F.: Deformation Structures in Glaciolacustrine Deposits of Khibiny and Assessment of Their Seismogenic Potential, Seismic Instruments, 53, 224–233, 2017.

Hestnes, E.: Slushflow hazard – where, why and when? 25 years of experience with slushflow consulting and research, Ann. Glaciol., 26, 370–376, 1998.

Hestnes, E. and Kristensen, K.: The diversities of large slushflows illustrated by selected cases, in: Proceedings of the International Snow Science Workshop, 17–22 October 2010, Squaw Valley, CA, USA, 348–355, 2010.

Ivanov, M. M., Belyaev, V. R., Garankina, E. V., Gurinov, A. L., Dobrydneva, L. V., Kuzmenkova, N. V., Romanenko, F. A., and Tulyakov, E. D.: Application of ^{232}Th Daughter Radionuclides for Studying Exogenic Processes, Izvestiya of Russian Academy of Sciences, Geography Series, 6, 81–92, 2018 (in Russian).

Ivanov, M. M., Kuzmenkova, N. V., Garankina, E. V., and Tulyakov, E. D.: The ^{232}Th distribution in modern sediments near radioactive lovchorrite mine, the Khibiny Mountains, Kola Peninsula, J. Geochem. Explor., 200, 77–83, https://doi.org/10.1016/j.gexplo.2019.01.005, 2019.

Jonsson, A. and Gauer, P.: Optimizing mitigation measures against slush flows by means of numerical modelling – a case study Longyearbyen, Svalbard, in: Proceedings of INTERPRAEVENT-2014 in the Pacific Rim – Natural Disaster Mitigation to Establish Society with the Resilience, 25–28 November 2014, Nara, Japan, 224–225, 2014.

Kobayashi, S., Izumi, K., and Kamiishi, I.: Slushflow disasters in Japan and its characteristics, International Snow Science Workshop, 30 October–3 November 1994, Snowbird, Utah, USA, 657–665, 1994.

Kosareva, Y. M.: Soils evolution in high mountains of Khibiny in Holocene, PhD Thesis, Moscow, Russia, 2007 (in Russian).

Larocque, S. J., Hetu, B., and Filion, L.: Geomorphic and Dendroecological impacts of slushflows in Central Gaspe Peninsula (Quebec, Canada), Geogr. Ann. A, 83, 191–201, 2001.

Nikolaeva, S. B.: Fold deformations in the Late Pleistocene deposits of Khibiny (the central part of the Kola Peninsula): morphology and genesis, Vestnik of Moscow State Technical University, 17, 329–339, 2014 (in Russian).

Nikolaeva, S. B., Lavrova, N. B., Denisov, D. B., and Tolstobrov, D. S.: Traces of catastrophic processes in bottom sediments of lakes on the western coast of Lake Babinskaya Imandra, Izvestiya of Russian Geographic Society, 148, 38–52, 2016 (in Russian).

Nyberg, R.: Observation of slushflows and their geomorphical effects in the Swedish mountain area, Geogr. Ann., A, 71, 185–198, 1989.

Perov, V. F.: Debris flows in Khibiny mountain massif, Vestn. Mosk. U., Seriya 5 Geographicheskaya, 1, 106–110, 1966 (in Russian).

Perov, V. F.: Attempt to use dendrochronological method for studying frequency of debris flows at Khibiny, in: Fitoindikatsionnyye metody v glyatsiologii, MSU Publ., Moscow, Russia, 42–49, 1971 (in Russian).

Perov, V. F.: Debris flow phenomena: terminological dictionary, MSU Pupbl., Moscow, Russia, 1996 (in Russian).

Perov, V. F., Chernomorets, S., Budarina, O., Savernyuk, E., and Leontyeva, T.: Debris flow hazards for mountain regions of Russia: regional features and key events, Nat. Hazards, 88, 199–235, https://doi.org/10.1007/s11069-017-2841-3, 2017.

Pozhilenko, V. I., Gavrilenko, B. V., Zhirov, B. B., and Zhabin, S. V.: Geology of ore zones of Murmansk region, KNC RAS Publ., Apatity, Russia, 2002 (in Russian).

Relf, G., Kendra, J., Schwartz, R., Leathers, D., and Levia, D.: Slushflows: science and planning considerations for an expanding hazard, Nat. Hazards, 78, 333–354, https://doi.org/10.1007/s11069-015-1716-8, 2015.

Romanenko, F. A., Lukashov, A. A., and Shilova, O. S.: Catastrophic gravitational processes at the North of European Russia and an attempt of their absolute age dating, Geomorfologiya, 1, 87–94, 2011 (in Russian).

Rudinskaya, A. I., Belyaev, Y. R., Gurinov, A. L., and Garankina, E. V.: Debris flow phenomena in the Lovozerskiye Tundry, in: Debris Flows: Disasters, Risk, Forecast, Protection, Proceedings of the 5th International Conference, 1–5 October 2018, Tbilisi, Georgia, 529–537, 2018 (in Russian).

Sapunov, V. N.: Slushflows in Khibiny, Materialy gljaciologicheskih issledovanij, 71, 94–99, 1991 (in Russian).

Shilova, O. S. and Romanenko, F. A.: Tracking of Holocene evolutionary and catastrophic processes in lake sediments of Schuchje Lake (Khibiny, Kola Peninsula), in: Proceedings of the All-Russian Scientific Conference, dedicated to the memory of Professor A. A. Velichko, 23–25 November 2016, Moscow, Russia, 1, 381–385, 2016 (in Russian).

Sulerzhitskiy, L. D., Ryabinin, A. O., Zaychuk G. I., and Vinogradova, S. N.: Radiocarbon dates of Geological Institute Laboratory of the AS USSR, Byulleten komissii po izucheniyu chetvertichnogo perioda, 5, 145–153, 1986 (in Russian).

Vashchalova, Ò. V.: Paleogeographic approach to reconstruct snow avalanche activity for long-term forecasting by example of Khibiny, in: Assessment and long-term forecast of environmental changes of mountains, MSU Publ., Moscow, Russia, 120–128, 1987 (in Russian).

Vladychensky, A. C., Kovaleva, N. O., Kosareva, Y. M., and Ivanov, V. V.: Soils Mineralogy and Bulk Composition in Trough Valleys of Khibiny Mountain Range, Doklady po ekologicheskomu pochvovedeniyu, 1, 1–19, 2007 (in Russian).

Proc. IAHS, 381, 49–54, 2019
https://doi.org/10.5194/piahs-381-49-2019

Initial results of sediment yield measurement interpretation using a regional approach: Southern Italy case study

Paolo Porto[1,2] **and Giovanni Callegari**[2]

[1]Department of Agraria, University Mediterranea of Reggio Calabria, Reggio Calabria, 89122, Italy
[2]CNR – Istituto per I Sistemi Agrari e Forestali per il Mediterraneo,
Sezione Ecologia e Idrologia Forestale, Rende (Cs), Italy

Correspondence: Paolo Porto (paolo.porto@unirc.it)

Published: 1 August 2019

Abstract. The theoretical TCEV (Two Component Extreme Value) distribution was applied to interpret the sediment yield datasets available in Southern Italy. The analysis is based on hydrological data collected for twelve catchments located in Calabria and Basilicata. A hierarchical approach was used to obtain a regional parent distribution which was used to determine the return time for each event. The hierarchical approach proposed in this study includes two stages. The first stage served for calibration and made it possible to estimate the parameters of the theoretical TCEV distribution. More specifically, the hypothesis of homogeneity with regard to the skewness coefficient and the coefficient of variation was verified using the datasets related to nine catchments. The second stage consisted in verifying the goodness of the theoretical distribution on three independent datasets provided by three experimental catchments not involved in the calibration. Overall results show that, even if the TCEV distribution was conceived to estimate peak flow, its concept of "double component" can be extended to predict sediment yield on a regional scale.

1 Introduction

Prediction of soil erosion and sediment yield, corresponding to a fixed return time, is necessary to adopt soil conservation strategies for reducing costs of on-site and off-site impacts in areas at risk.

In Southern Italy, high rates of soil erosion and sediment yield are well documented both on cultivated land (Porto and Walling, 2012) and in areas covered by forests (Porto et al., 2004). Over the last 60 years, predictions of sediment yield are based both on the use of traditional monitoring techniques, such as observations on experimental plots or catchments, and on different types of models that vary from empirical-parametric approaches, such as SEDD (Ferro and Porto, 2000) or revised versions of the USLE (Cinnirella et al., 1998) to more recent physically- based models, such as WEPP (Nearing et al., 1989), whose goal is to simulate both the detachment and transport of soil particles.

However, even if these models provide good estimates of long-term average values of sediment yield, they cannot make reliable predictions associated to a fixed return time especially if they are associated with extreme events.

A recent study (Porto and Callegari, 2019), carried out on three experimental catchments (W1, W2, and W3) located in Calabria (Southern Italy), demonstrated that a significant proportion of the annual sediment yield in these catchments is due to the extreme event occurred each year which produced more than 50 % of the total annual sediment yield with a very few exceptions.

Looking at the frequency distribution of these measurements of sediment yield, the same authors showed that the empirical distribution of the normalized variable $x_i = \mathrm{SY}_{\mathrm{MAX},i}/\mu(\mathrm{SY})$, where $\mathrm{SY}_{\mathrm{MAX},i}$. is the maximum annual value and $\mu(\mathrm{SY})$ is the mean annual value of the entire monitoring period, is skewed. In particular, results of observation on the catchment W2 (see Porto and Callegari, 2019) shows that two components can be identified (Fig. 1). One of them

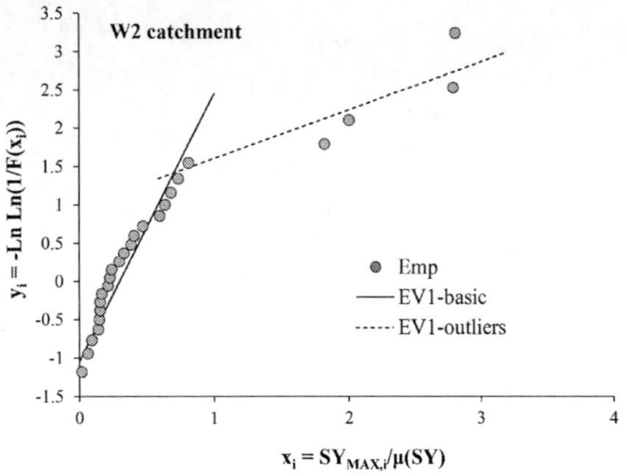

Figure 1. Gumbel's plot for the dataset of the normalized variable, $x_i = \mathrm{SY}_{\mathrm{MAX},i}/\mu(\mathrm{SY})$.

is associated with ordinary events (basic component) and a second one is characterized by extreme events (outliers component).

The authors concluded that the process can be represented by two Extreme Value Type 1 (EV1) independent distributions, one for each component, capable of associating each event with a corresponding return time T. The double component indicated in Fig. 1 is a well known effect when studying annual records of maximum peak flow values (see Rossi et al., 1984) and a more sophisticated approach can be used to explain it.

This approach is based on the TCEV (Two Component Extreme Value) model whose ability was well demonstrated in predicting peak flow and critical rainfall corresponding to a fixed return time (see Rossi et al., 1984; Ferro and Porto, 1999, 2006). The TCEV distribution is based on the assumption that in each historical sample of the observed variable x, one or several annual maximum values are significantly higher than the bulk of the remaining data, so that two components are clearly distinguishable.

Therefore, the theoretical CDF of the TCEV law $F(x)$ can be expressed as a probabilistic model with four parameters derived from the product of two EV1 distributions:

$$F(x) = \exp\left[-\lambda_1 \exp\left(-\frac{x}{\theta_1}\right) - \lambda_2 \exp\left(-\frac{x}{\theta_2}\right)\right] \quad (1)$$

where: λ_1, λ_2 are the shape parameters and represent the mean number of events which belong respectively, to the basic (component 1) and the outlying (component 2) components; and θ_1, θ_2 are the scale parameters, which represent the at-site central values of the x variable for each component.

The basic component is characterized by a high number of events and by values of x less than those corresponding to the outlying component so that $\lambda_1 \gg \lambda_2$ and $\theta_1 \ll \theta_2$. The main advantage of the TCEV model, if compared to the EV1 distribution, is based on the fact that the former can be used at a regional scale because their parameters (or their combination) showed to be scale invariant for large areas.

In order to apply the model at a regional scale it is useful to introduce the standardized variable $y = \left(\frac{x}{\theta_1}\right) - \lambda_1$ by which Eq. (1) becomes:

$$F(y) = \exp\left[-\exp(-y) - \Lambda_* \exp\left(-\frac{y}{\Theta_*}\right)\right] \quad (2)$$

where: $\Theta_* = \theta_2/\theta_1$ and $\Lambda_* = \lambda_2/\lambda_1^{1/\Theta_*}$.

Introducing the dimensionless variable x', equal to the ratio between x and the mean value μ of the TCEV distribution, Eq. (2) becomes:

$$F(x') = \exp\left\{-\lambda_1[\exp(\alpha)]^{-x'} - \Lambda_*\lambda_1^{1/\Theta_*}\left[\exp\left(\frac{\alpha}{\Theta_*}\right)\right]^{-x'}\right\} \quad (3)$$

named growth curve, in which:

$$\alpha = \frac{\mu}{\theta_1} = 0.5772 + \ln\lambda_1 - \sum_{j=1}^{\infty} \frac{(-1)^j \Lambda_*^j \Gamma\left(\frac{j}{\Theta_*}\right)}{j!} \quad (4)$$

where $\Gamma()$ is the gamma function.

For fixed return period T, Eq. (3) requires only 3 parameters (Λ_*, Θ_* and λ_1) for the calculation of the variable x' into a selected area (i.e. conversion of growth curve to quantile). It is assumed that the first two parameters, Λ_* and Θ_*, are constant over a very large area and the parameter λ_1 is considered constant for each sub-region into which the region is divided. When the three parameters of Eq. (3) are defined and taking into account of the definition of the dimensionless variable x', the corresponding value of the variable x can be obtained by the following equation:

$$x = x'\mu \quad (5)$$

where: the variable μ (or scale parameter) can be replaced by the empirical mean of the single records.

Considering the logical dependence of sediment yield on the magnitude of peak flow (see among the others Williams and Berndt, 1977), the aim of this work is to interpret the observed values of sediment yield using an approach based on the TCEV distribution.

2 Materials and methods

The Italian Hydrographic Service (SIMI) has published long-term measurements of daily and monthly suspended sediment load (SSL) on many catchments throughout the country. In the present contribution, monthly measurements of SSL, which were collected on 11 rivers located in Calabria, and Basilicata, have been used for analysis. This step required also a preliminary aggregation of the data to generate

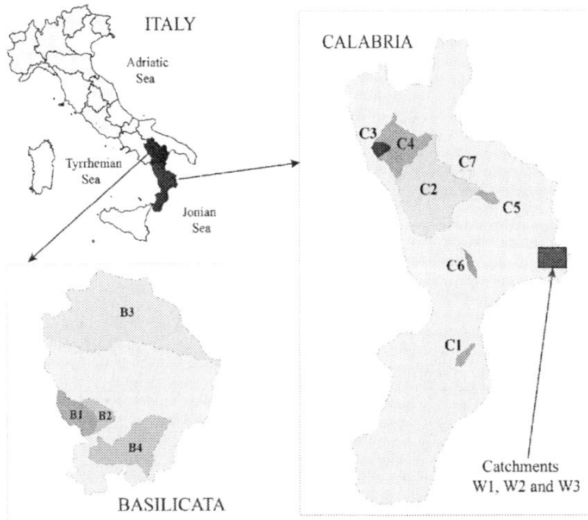

Figure 2. Location of the catchments investigated in this study.

Table 1. Details of stream gauges and rivers considered in this study.

River	Station	Code	Years	N	Area (km^2)
BASILICATA					
AGRI	Grumento	B1	1962–1972	10	278.0
AGRI	Tarangelo	B2	1968–1972	5	507.0
BRADANO	Tavole Palatine	B3	1933–1971	34	2743.0
SINNI	Valsinni	B4	1938–1973	26	1142.0
CALABRIA					
ALACO	Mammone	C1	1961–1979	18	14.8
CRATI	Conca	C2	1933–1970	31	1332.0
ESARO	Cameli	C3	1964–1971	8	55.4
ESARO	La Musica	C4	1960–1969	9	532.0
LESE	Schiena d'Asino	C5	1953–1980	19	60.0
MELITO	Olivella	C6	1962–1977	14	41.2
TRIONTO	Difesa	C7	1964–1979	15	31.7

the annual datasets of sediment yield (see Table 1 and Fig. 2 for details).

These datasets have been integrated by 3 long-term records of sediment yield obtained in three experimental catchments (W1, W2, and W3) located in Calabria (see Porto et al., 2004). In this case, the original measurements, collected at event scale, have been aggregated at annual scale to make them comparable with the previous datasets.

More specifically, the 11 datasets were used to derive the growth curve (Eq. 3) for the entire region, here assumed as a unique homogeneous area, while the three catchments W1, W2, and W3 were used to test the validity of the Eq. (3) on independent datasets located in the same region.

In other words, the hierarchical approach proposed in this study consists of two levels: (1) the first level aimed at calculating the 3 parameters (Λ_*, Θ_* and λ_1) of the Eq. (3) on the area where the catchments are located. In this respect, a comparison between the empirical and theoretical CDFs of γ was performed; (2) the second level served to validate the Eq. (3) on the three independent datasets for which the empirical CDFs of the variable x are available.

3 Results and discussions

3.1 Model calibration

The analysis carried out at the first level required the basic assumption that the entire region can be considered homogeneous with respect to the skewness coefficient, γ and the coefficient of variation, CV (see Fiorentino et al., 1987; Beran et al., 1986). However, considering the short length of some historical datasets ($N < 10$ years), some preliminary adjustments of these records were necessary. More specifically, based on geomorphic similitude, the two datasets of River Agri (see Table 1) were bulked into one single dataset with $N = 15$; the same assumption was made for the two records of River Esaro for which a final single dataset with $N = 17$ was obtained.

As explained above, in the first step of this analysis, the TCEV model required the estimation of the regional parameters Λ_* and Θ_* referred to these new nine series. This analysis was conducted using the procedure suggested by Fiorentino and Gabriele (1985) that led to the following ML estimates of the two parameters:

$$\Lambda_* = 0.675; \quad \Theta_* = 11.574 \tag{6}$$

In order to verify the homogeneity of the entire region with regard to γ, Monte Carlo experiments were performed for generating theoretical sequences having a fixed size N and distributed according to a TCEV model with regional parameters given by Eq. (6).

The CDF of the empirical skewness related to the 9 samples and that generated by the regional model are compared in Fig. 3. The theoretical CDF was obtained using a Monte Carlo technique based on 20 000 synthetic series having a sample size N equal to the mean value calculated from the empirical datasets.

Even if a perfect overlapping cannot be expected considering the limited number of datasets, Fig. 3 shows a certain ability of the TCEV distribution to reproduce the empirical skewness CDF.

The second step of the analysis is based on the assumption that the investigated region should be divided into smaller homogeneous areas, named sub-regions, in which the λ_1 parameter can be assumed constant. Because the λ_1 parameter is related to CV by the following equation:

$$\mathrm{CV} = \frac{0.577}{\log \Lambda_1 + 0.251} \tag{7}$$

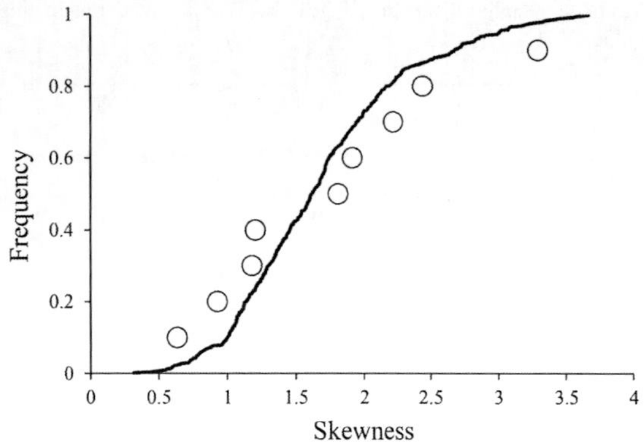

Figure 3. Comparison between the empirical CDF of the skewness coefficient calculated from the nine datasets and the regional CDF obtained from the theoretical distribution.

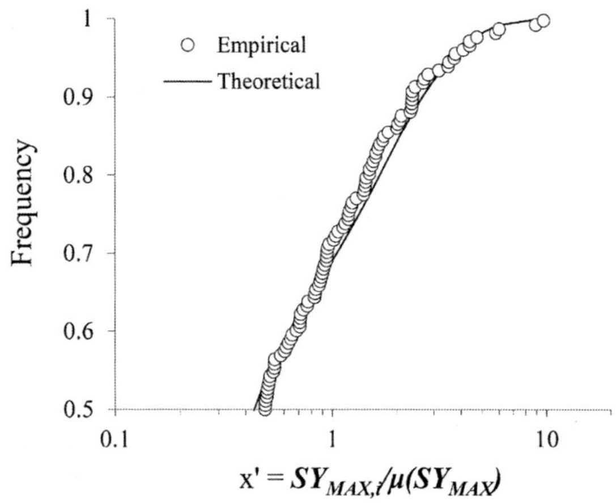

Figure 4. Comparison between empirical and theoretical CDF of the variable x' for the 9 investigated datasets.

this is equivalent to defining sub-regions within which the coefficient of variation of the individual series can be assumed constant. Unfortunately, the number of series available in the entire region (9 datasets) is not enough to proceed with a further geographic subdivision and this hypothesis could not be considered. In the light of this limitation, the analysis was carried out considering a unique homogeneous region even at the second hierarchical level. Consequently, according to Arnell and Gabriele (1988), the value of Λ_1 was determined using the TCEV distribution with the regional values of Λ_* and Θ_* given by Eq. (6). Using the data obtained from the historical datasets, the parameters Λ_1 and α assume the following values:

$$\Lambda_1 = 5.974; \; \alpha = 8.776 \tag{8}$$

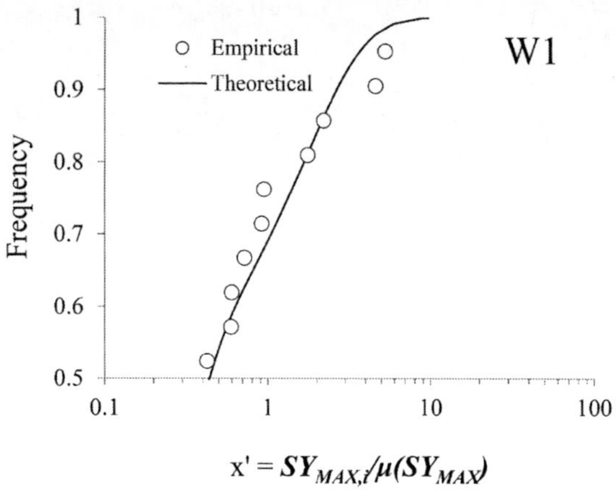

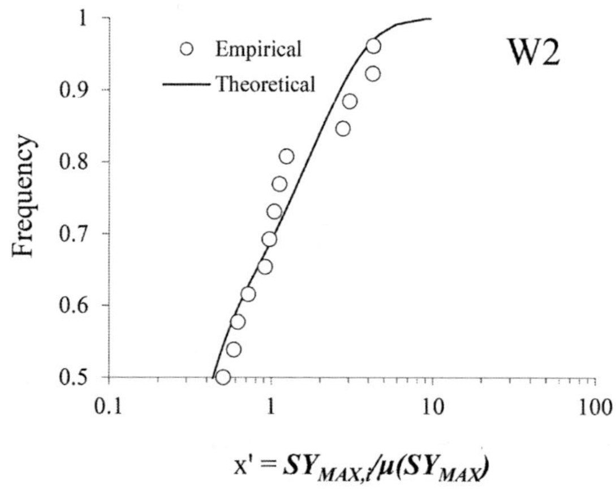

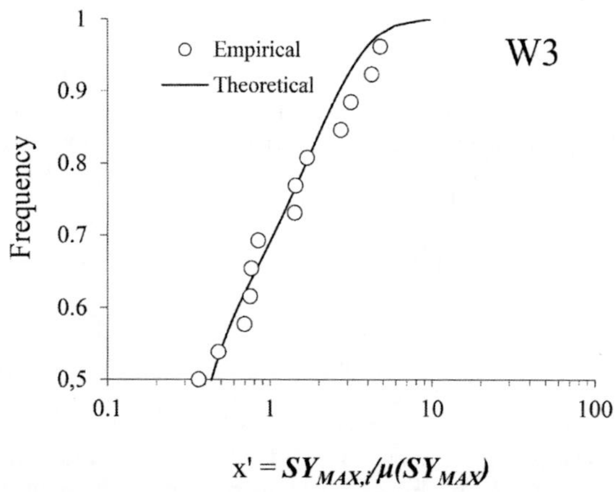

Figure 5. Comparison between empirical and theoretical CDFs of the variable x' for the 3 independent datasets (W1, W2 and W3).

Figure 4 shows that the growth curve calculated by the Eq. (3), with the parameters given by the Eqs. (6) and (8) fits very well the higher values ($f(x') > 0.5$) of the x' empirical distribution suggesting that the hypothesis of unique homogeneous region is plausible.

3.2 Model performance

The regional growth curve given by the Eq. (3) with the calibrated values of the 4 parameters (Λ_*, Θ_* , λ_1 and α) was tested on three independent datasets available for three experimental catchments (W1, W2, and W3) located in Calabria (see Fig. 2). Details of these catchments are reported elsewhere (see Ferro and Porto, 2000). Here, only the CDFs of the empirical variable x' are considered and the comparison with the theoretical distribution is reported for each catchment in Fig. 5.

The visual inspection of the three graphs in Fig. 5 suggests that, the model fits reasonably well the higher values ($f(x') > 0.5$) of the x' empirical distributions suggesting that the regional growth curve is very robust and can be used for obtaining reliable predictions even for large events.

4 Conclusions

Prediction of sediment yield generated by large events is very important in areas not covered by monitoring stations or for the areas with limited observational periods. Based on a previous contribution by Porto and Callegari (2019), the study presented herein reports a regional investigation carried out in Southern Italy aimed at evaluating the frequency distribution of sediment yield. The procedure applied in this contribution is based on the use of the TCEV model and its hierarchical approach to get reliable estimates of the regional parameters of the distribution. The choice is due to the ability of the TCEV distribution to reproduce the skewed empirical distributions of sediment yield observed in many historical datasets and to generate estimates of sediment yield associated with a fixed return time. Overall, the study demonstrated that even if the TCEV distribution was conceived to provide estimates of peak flow, its basic assumptions and its hierarchical procedure can be used to predict sediment yield at regional scale.

Data availability. Data collected by the Italian Hydrographic Service (SIMI) can be downloaded from the website http://www.gruppoalluvioni.it/annali-idrologici/ (last access: 1 August 2018). Data related to the catchments W1, W2 and W3 are available upon request to the authors.

Author contributions. All authors setup the research, analyzed the results and participated in writing the paper.

Competing interests. The authors declare that they have no conflict of interest.

Special issue statement. This article is part of the special issue "Land use and climate change impacts on erosion and sediment transport". It is a result of the ICCE Symposium 2018 – Climate Change Impacts on Sediment Dynamics: Measurement, Modelling and Management, Moscow, Russia, 27–31 August 2018.

Acknowledgements. The study has been finalized in the frame of the Erasmus + KA2 – Cooperation for innovation and the exchange of good practices – Capacity Building in the field of Higher Education – Soil Erosion and Torrential Flood Prevention: Curriculum Development at the Universities of Western Balkan Countries/SETOF.

Financial support. This research has been supported by the European Project SETOF (grant no. 598403-EPP-1-2018-1-RS-EPPKA2-CBHE-JP).

References

Arnell, N. W. and Gabriele, S.: The performance of the two component extreme value distribution in regional flood frequency analysis, Water Resour. Res., 24, 879–887, 1988.

Beran, M., Hosking, J. R. M., and Arnell, N. W.: Comment on 'Two component extreme value distribution for flood frequency analysis', by Rossi et al., Water Resour. Res., 22, 263–266, 1986.

Cinnirella, S., Iovino, F., Porto, P., and Ferro, V.: Antierosive effectiveness of Eucalyptus coppices through the cover management factor estimate, Hydrol. Process. 12, 635–649, 1998.

Ferro, V. and Porto, P.: Regional analysis of rainfall-depth-duration equation for South Italy, J. Hydrol. Eng., 4, 326–336, 1999.

Ferro, V. and Porto, P.: Sediment Delivery Distributed (SEDD) Model, J. Hydrol. Eng., 5, 411–422, 2000.

Ferro, V. and Porto, P.: Floods frequency analysis for Sicily (South Italy), J. Hydrol. Eng., 11, 110–122, 2006.

Fiorentino, M. and Gabriele, S.: Distribuzione TCEV: metodi di stima dei parametri e proprietà statistiche degli stimatori, Geodata n. 25, Cosenza, 1985 (in Italian).

Fiorentino, M., Gabriele, S., Rossi, F., and Versace, P.: Hierarchical approach for regional flood frequency analysis. Regional Flood Frequency Analysis, edited by: Singh, V. P., D. Reidel Publishing Company, Springer, the Netherlands, 2, 39–49, 1987.

Nearing, M. A., Foster, G. R., Lane, L. J., and Finkner, S. C.: A process-based soil erosion model for USDA-Water Erosion Prediction Project Technology, Transactions of the ASAE, 32, 1587–1593, 1989.

Porto, P. and Callegari, G.: A Probabilistic Approach to Interpret Long-Term Observations of Sediment Yield in Experimental Catchments in Southern Italy, in: Climate Change Impacts on Sediment Dynamics: Measurement, Modelling and Management, edited by: Chalov, S., Golosov, V., Li, R., and Tsyplenkov, A., Springer Nature Switzerland AG, Cham, Switzerland, 110–117, 2019.

Porto, P. and Walling, D. E.: Using plot experiments to test the validity of mass balance models employed to estimate soil redistribution rates from ^{137}Cs and $^{210}Pb_{ex}$ measurements, Appl. Radiat. Isotopes, 70, 2451–2459, 2012.

Porto, P., Walling, D. E., and Callegari, G.: Validating the use of caesium-137 measurements to estimate erosion rates in three small catchments in Southern Italy, IAHS Publ., 288, 75–83, 2004.

Rossi, F., Fiorentino, M., and Versace, P.: Two-component extreme value distribution for flood frequency analysis, Water Resour. Res., 20, 847–856, 1984.

Williams, J. R. and Berndt, H. D.: Sediment Yield Prediction Based on Watershed Hydrology, Transactions of the ASAE, 20, 1100–1104, 1977.

Proc. IAHS, 381, 55–64, 2019
https://doi.org/10.5194/piahs-381-55-2019

Suspended sediment yield and climate change in Kamchatka, Far East of Russia

Ludmila V. Kuksina

Laboratory of soil erosion and channel processes, Faculty of geography,
Moscow State University, Moscow, 119991, Russia

Correspondence: Ludmila V. Kuksina (ludmilakuksina@gmail.com)

Published: 1 August 2019

Abstract. Key factors of sediment yield formation and variability in the Kamchatsky Krai are relief characteristics and climate (especially precipitation). The last research in climate change in Kamchatka demonstrates climate warming – mean annual air temperature has increased on 1.5 °C during the period from 1951 to 2009 in average (the maximum increase was revealed in the west of Kamchatka and in the Kamchatka River valley, the minimum one – in the extreme south and north). Decrease of annual precipitation was observed at the same period. In addition frequency and intensity of precipitation is growing. Thereby described climate changes in Kamchatka influence on sediment yield characteristics directly (influence of precipitation on surface washout formation) and indirectly (through water runoff and hydrological regime of rivers changes). Analysis of spatio-temporal variability in suspended sediment yield demonstrated decrease since late 1970–early 1980 in the most part of rivers in the Kamchatsky Krai. Revealed fluctuations demonstrate very good coincidence with precipitation amount variability. Decrease of precipitation leads to corresponding diminution of rainfall erosive factor in the territory which is the key factor of suspended sediment yield formation and variability in rivers in the Kamchatsky Krai. The value of reduction varies from 1 % to 45 % and involves decrease of erosion rates in the territory. Thereby climate change influences on suspended sediment yield in multiple-valued way. Income of friable volcanic deposits on the catchment surface is the key factor of suspended sediment yield formation and variability in some river basins under the impact of volcanic eruptions, and in combination with raising precipitation it can lead to catastrophic increase of sediment yield.

1 Introduction

The current climate change occurs all over the World and leads to increase of air temperature, amount, intensity and frequency of precipitation, and rainfall erosive factor, and as a consequence to growth of severe floods and erosive events (landslides, mudflows (including lahars)) causing the increase of sediment yield, basin component of sediment yield, and channel transformations. According to the last Report on climatic risks in the Russian Federation (Report on climate risk, 2017) climate changes are faster in Russia in comparison to the Earth, and severity of hydrological events and their damage growths correspondingly. Increase of high precipitation frequency (Groisman et al., 2005) and warming climate (Huntington, 2006) leads to the growth of power and frequency of floods, relief reformation and high sediment yields (Kuksina et al., 2017; Mueller, Pfister, 2011), and these extreme hydrometeorological events rank the first position among global risks threatening humanity (Report on climate risk, 2017).

The Kamchatsky Krai could be very vulnerable to climate change due to natural features of the territory including vast distribution of friable volcanic deposits which are very easy-portable by water (Kraevaya, 1969), and specific exposed landscapes (repetitive catastrophic landslides in Geyser Valley in 2007, 2014, and 2017 due to unstable rocks on its flanks, Leonov et al., 2014). Volcanic activity is specific feature of sediment yield formation and spatio-temporal distribution in Kamchatka (Kuksina and Alexeevsky, 2017). Volcanic eruptions are the first-rate source of sediments in planetary scale. Among the whole natural phenomena leading to extreme increase of sediment yield (volcanic eruptions, earthquakes, heavy rains, and dam failure), volcanoes are the absolute leaders (Korup, 2012). Highly erodible volcanogenic deposits fall in the river basins during volcanic eruptions due to ashfalls, pyroclastic flows, and lahars (Carrivick et al., 2010; Gran and Montgomery, 2005; Lavigne and Thouret, 2002; Major, 2004).

Volcanic eruptions lead to extreme increasing of suspended sediment yield in rivers draining flanks of volcanoes. Annual suspended sediment yield exceeded background level in 500 times after the catastrophic eruption of Mount St Helens in 1980. Volcanic eruptions impact on hydrological regime of rivers. Within 5–10 years after the catastrophic eruption of Mount St Helens on the 18 May 1980 increasing of peak discharges was observed in autumn and winter due to decrease of lag-time and increase of channel efficiency (Major and Mark, 2006). In some cases the processes of water transfer changes in comparison with the period before eruption. In the post eruption period loss of canopy interception, the greater amount of precipitation reaching the ground surface, reduced evapotranspiration, the reduction of surface infiltration and more prominent overland flow were observed (Major et al., 2009). Thereby volcanic activity could be the separate reason of water runoff and sediment yield fluctuations in Kamchatka rivers, and it can influence on river runoff characteristics through increasing precipitation. Combination of such events as high precipitation, increase of air temperature and intensification of snow and ice melt, and volcanic eruptions could lead to catastrophic natural events and great increase of sediment yield (lahars, lake burst, etc.) (Korup, 2012). Most part of the rivers in Kamchatka is spawning area for valuable salmon species (such as king salmon, red salmon, silver salmon, etc.), and for some of them catch of fish exceeds 90 % of the total amount in the Russian Far East. River ecosystems and spawning areas depend on hydrological conditions including suspended sediment concentration.

The last research in climate change in Kamchatka (Shkaberda and Vasilevskaya, 2014) demonstrates climate warming – mean annual air temperature has increased on 1.5 °C during the period from 1951 to 2009 in average (the maximum increase was revealed in the west of Kamchatka and in the Kamchatka River valley, the minimum one – in the extreme south and north); "continentality" of climate decrease on 6 %–16 % in some regions of the peninsula. Variability of annual precipitation was multiple-valued at the same period. Decrease of precipitation was observed at the same period almost everywhere in the Kamchatsky Krai. In addition frequency and intensity of precipitation is growing (Groisman et al., 2005). Climate change leads to correspondence variations in water runoff and sediment yield in Kamchatka.

Key factors of sediment yield formation and variability in the Kamchatsky Krai are relief characteristics and precipitation (Kuksina and Alexeevsky, 2018). Thereby described climate changes in Kamchatka influence on sediment yield characteristics directly (influence of precipitation on surface washout formation) and indirectly (through water runoff and hydrological regime of rivers changes). Thereby climate change influences on suspended sediment yield in multiple-valued way. Income of friable volcanic deposits on the catchment surface is the key factor of suspended sediment yield formation and variability in some river basins under the impact of volcanic eruptions.

The main goal of this paper is analysis of the features of spatial and temporal variability of suspended sediment yield in the state-of-the-art (current) hydroclimatic conditions.

2 Materials and methods

This study is based on stationary monitoring data on air temperature, precipitation, water runoff and suspended sediment yield characteristics. The data on air temperature and precipitation contains information for 21 stations (Fig. 1), and duration of observations varies from 80 (Oktyabrskaya) to 174 years (Petropavlovsk-Kamchatsky) for air temperature, and from 16 to 51 for precipitation.

There were 269 monitoring stations that measured water runoff and levels in the territory of the Kamchatsky Krai for the period of hydrometeorological observations (since 1930). 86 gauges operated in 2010. The average duration of the observations is 22 years. The majority of the water runoff gauges are situated in river basins with areas less than 2000 km^2 (76 % of all gauges). The gauges are distributed very unevenly. There are a relatively large number of gauges in the large plain rivers (the Avacha River (12), the Kamchatka River (28)) that flow into the Pacific Ocean. There are only 8 operating gauges in the continental part of the Kamchatsky Krai now (6 of them are situated in the Penzhina River basin; 10 gauges are closed).

Regular monitoring of suspended sediment yield characteristics in rivers of the Kamchatsky Krai was started in 1940 in the Kamchatka River (gauge Kliuchi) and the Avacha River (gauge Elizovo). Between 1941–2015, suspended sediment yield monitoring was implemented in 69 stations. 6 of them had only episodic data, and they were excluded from this study's analysis. The duration of observations vary from 6 to 74 years for 63 selected gauges (see Fig. 1). More than half of the gauges (54 %) were situated in river basins with catchment areas greater than 500 km^2; hence, 46 % were in river basins with areas less than 500 km^2. 20 stations with observations on suspended sediment yield currently operate in the territory. The duration of observations of these gauges varies from 37 to 74 years.

The representativeness (Helsel and Hirsch, 2002) of all of the series of observations was evaluated, and even measurement periods as short as 6 to 10 years appear to be representative (probably the observation data hit into the water- and sediment-average period with absence of significant hydrological events). As a result, these data were used for defining

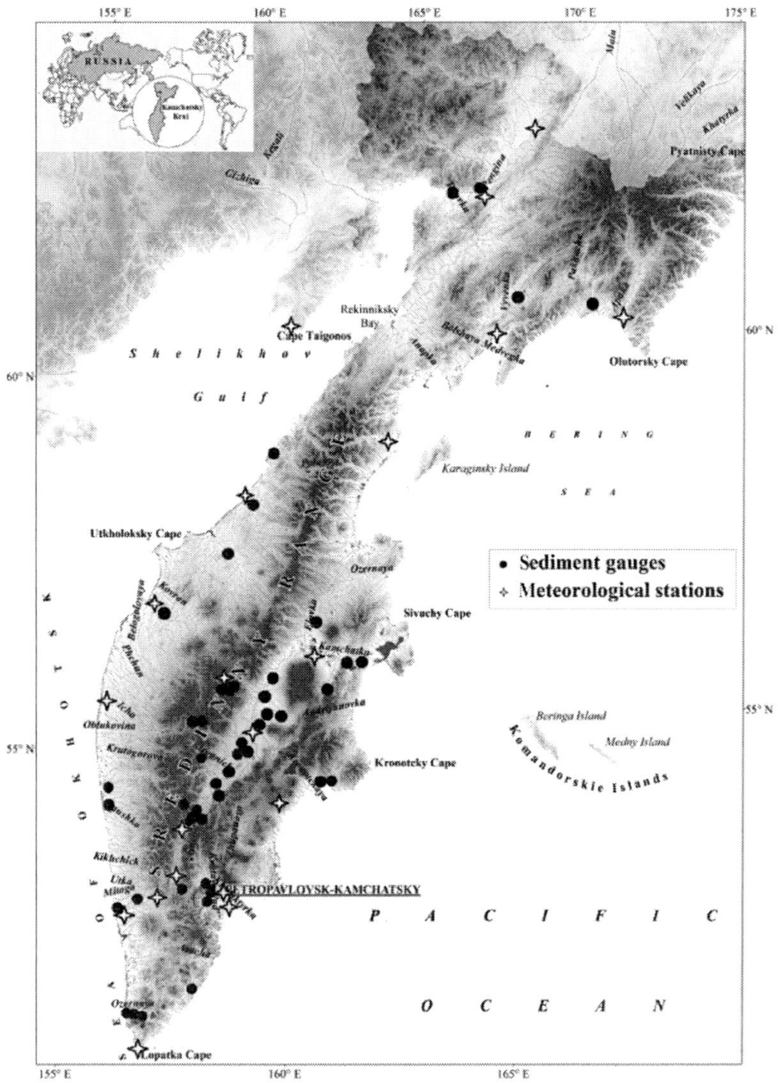

Figure 1. Distribution of suspended sediment yield and meteorological monitoring stations in the Kamchatsky Krai. Rivers: (1) Paren', (2) Pakhacha, (3) Penzhina, (4) Avacha, (5) Talovka, (6) Tigil', (7) Udova, (8) Asacha, (9) Bol'shaya Vorovskaya, (10) Bol'shaya (Bystraya), (11) Bryumka, (12) Phchyn, (13) Voyampolka, (14) Vyvenka, (15) Zhupanova, (16) Icha, (17) Kamchatka, (18) Kronotckaya, (19) Oblukovina, (20) Ozernaya, (21) Ozernaya, (22) Palana, (23) Khairyuzova.

water regime types and zones having similar suspended sediment yield characteristics.

The long-term trends of water runoff, suspended sediment yield, precipitation and air temperature variability were determined using a difference-integral curves analysis (Evstignnev, 1990; Manh et al., 2013).

3 Climate change in Kamchatka

The last research in climate change in Kamchatka (Shkaberda, 2014) demonstrates climate warming – mean annual air temperature has increased on 1.5 °C during the period from 1951 to 2009 in average (the maximum increase was revealed in the west of Kamchatka and in the Kamchatka River valley, the minimum one – in the extreme south and north); "continentality" of climate decrease on 6 %–16 % in

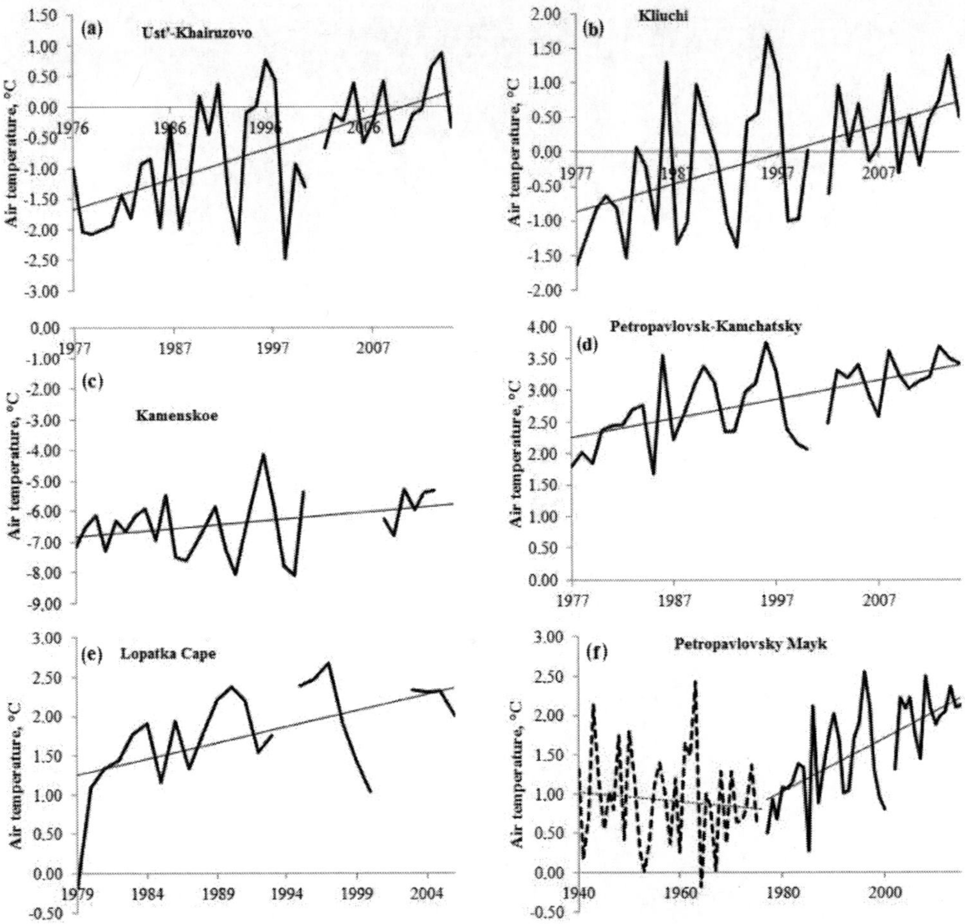

Figure 2. Trends in air temperature change in the western coast (Ust'-Khairuzovo station, **a**), central part (Kliuchu station , **b**), northern (continental) part (Kamenskoe station, **c**), eastern coast (Petropavlovsk-Kamchatsky station, **d**), Petropavlovky Mayk station, **(f)**, and the very south of the peninsula (Lopatka Cape station, **e**), in the Kamchatsky Krai.

some regions of the peninsula. Variability of annual precipitation was multiple-valued at the same period. Decrease of precipitation was observed in the north, western and north-eastern coast, and in the Kamchatka River valley. Increase of them was observed in the eastern coast and in mountainous regions.

Analysis of climate change in Kamchatka was implemented on the basis of annual air temperature and total annual precipitation fluctuations in 21 stations in the Kamchatsky Krai.

Analysis of all temperature ranges revealed intensive increase of that characteristic all over the Kamchatsky Krai. The mean annual air temperature has increased on 1.5 °C during the period from 1951 to 2009 in average (Shkaberda, 2014). The maximum increase was revealed in the western coast of the peninsula. In Ust'-Khairuzovo station the air temperature increased on 1.9 °C for the period from late 1970 to 2015 (Fig. 2a). Intensive growth in air temperature is also observed in the Kamchatka River valley. According to the Dolinovka and the Kluichi stations the temperature increased on 1.6 °C for the period from late 1970 to 2015 (Fig. 2b). Less increase of air temperature is registered in the north continental regions (not more than 1.0 °C), in the eastern coast of the peninsula (1.1 °C) and in the very south of the peninsula (Lopatka Cape, not more than 1.1 °C) (Fig. 2c, d, e).

The most intensive growth of air temperature has been observed since late 1970, and in some stations the tendency even changed to the opposite one. For example, air temperature decreased till the late 1970 in Kamenskoe, Petropavlovsky Mayak, Lopatka Cape and other stations, and then it started to increase rapidly (Fig. 2f).

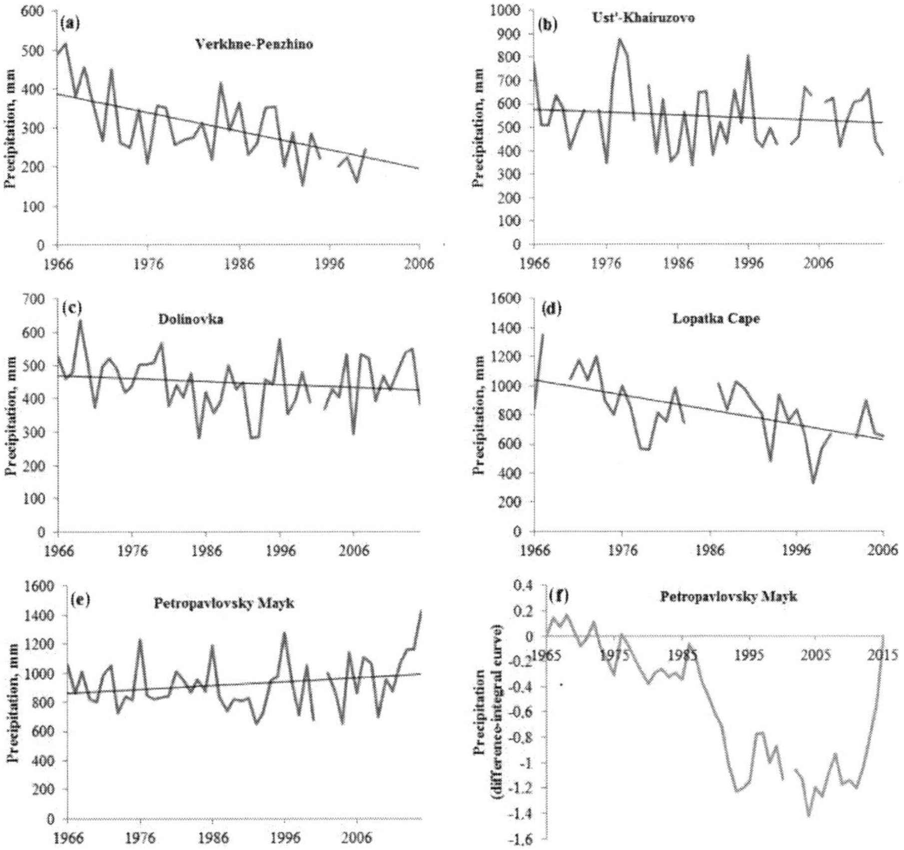

Figure 3. Trends in precipitation change in the northern (continental) part (Verkhne-Penzhino station, **a**), western coast (Ust'-Khairuzovo station, **b**), central part (Dolinovka station, **c**), the very south of the peninsula (Lopatka Cape station, **d**), and eastern coast (Petropavlovky Mayk station, **e, f**) in the Kamchatsky Krai.

Geographical position of the Kamchatsky Krai and orography of the region determine amount and distribution of the precipitation in the territory. The most intensive cyclonic activity is in the south; it subsides significantly to the north and decreases the mean annual amount of precipitation correspondingly. There are more precipitation on windward eastern flanks of mountains, while on leeward western flanks and protected river valleys. Analysis of precipitation data demonstrates variability of mean annual values from 300 mm yr^{-1} in the north of the region to 1310 mm yr^{-1} in the south-east of Kamchatka.

Variability of precipitation can influence on sediment yield directly (through intensity of sediment washout from the river basin territory) and indirectly (through variation of river water runoff). According to (Shkaberda, 2014) research the annual amount of precipitation varies in multiple-value way in Kamchatka. She mentioned decrease of precipitation was observed in the north, western and north-eastern coast, and in the Kamchatka River valley. Increase of them was observed in the eastern coast and in mountainous regions.

Our estimations demonstrate decrease of precipitation all over the Kamchatsky Krai territory but intensity of the process varies (Fig. 3a, b, c, d). The only one exception is Petropavlovsky Mayak station. In (Shkaberda, 2014) the increase of precipitation in that case is explained by positive anomaly in precipitation according to difference-integral curve (see Fig. 3e, f).

Changes in precipitation amount influence on rainfall erosive factor R which has particular importance for controlling suspended sediment yield. Rrepresents the number of rainfall erosion index units, and characterizes the erosive potential of precipitation (Wischmeier et al., 1971). For the Kamchatsky Krai, the distribution of the rainfall factor was obtained by combining data from R-distribution maps (Larionov, 1993) with more recent precipitation data (since 2015). The intensity of erosion depends on rainfall erosive fac-

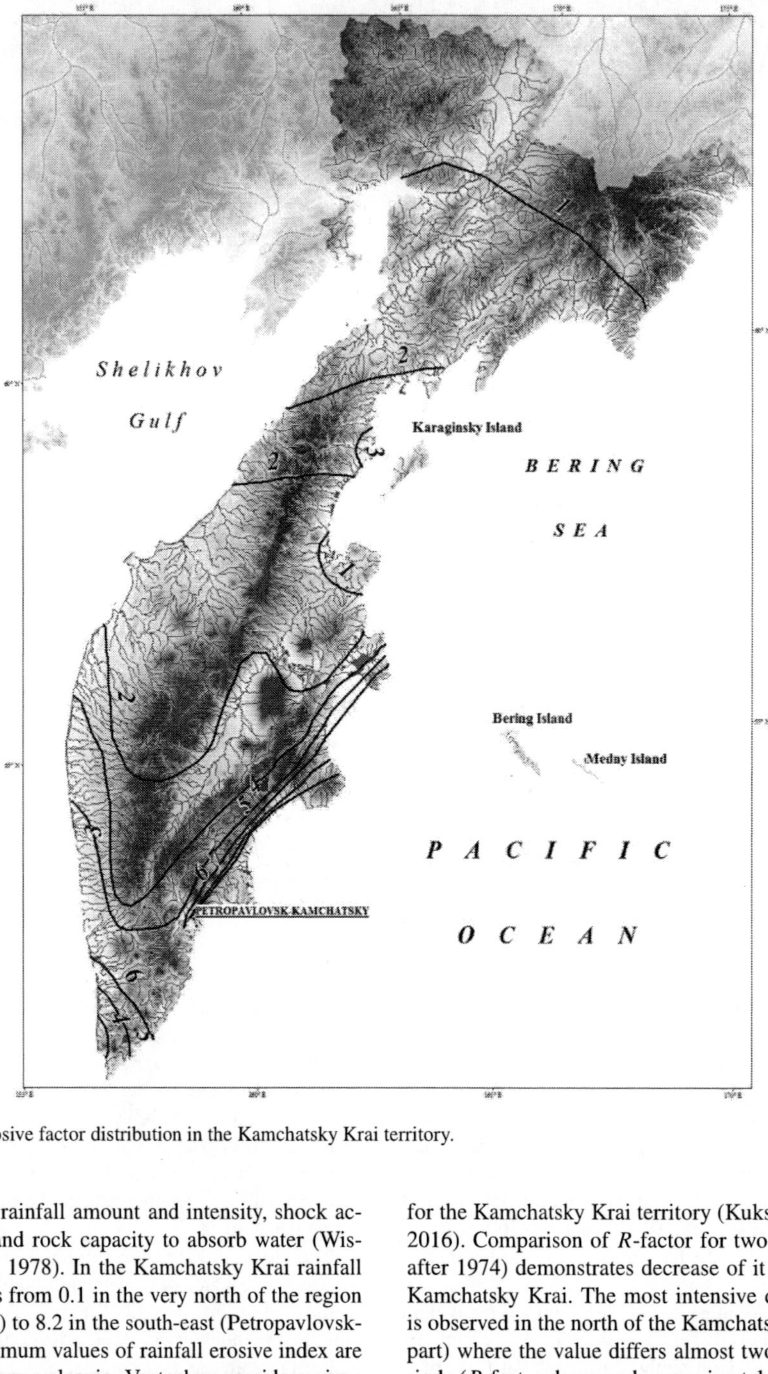

Figure 4. Rainfall erosive factor distribution in the Kamchatsky Krai territory.

tor considering the rainfall amount and intensity, shock action of rain drops and rock capacity to absorb water (Wischmeier and Smith, 1978). In the Kamchatsky Krai rainfall erosive factor varies from 0.1 in the very north of the region (Slautnoe settement) to 8.2 in the south-east (Petropavlovsk-Kamchatsky). Maximum values of rainfall erosive index are observed in Vostochny volcanic, Vostochny seaside regions and Zapadnaya Kamchatka plain. The minimum values characterizes the northern part of the region (Kuksina and Alexeevsky, 2014) (Fig. 4). It was found there is a positive correlation between R-factor and annual precipitation amount for the Kamchatsky Krai territory (Kuksina and Alexeevsky, 2016). Comparison of R-factor for two periods (before and after 1974) demonstrates decrease of it in all regions of the Kamchatsky Krai. The most intensive decrease of R-factor is observed in the north of the Kamchatsky Krai (continental part) where the value differs almost two times for these periods (R-factor decreased approximately in 45 %). Close results are observed in the very south of the peninsula (R-factor decreased approximately in 40 %). Decrease of R-factor in the west coast doesn't exceed 28 % (for Ust'-Kharuzova station it's 8 % only). In the eastern coast decrease of R-factor

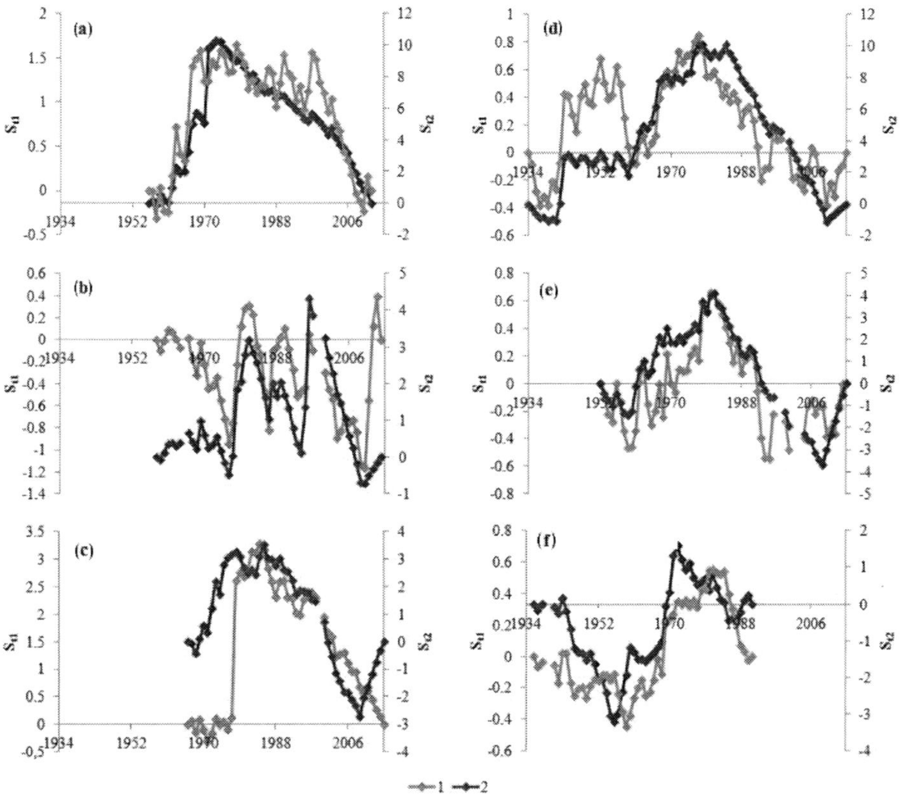

Figure 5. Trends in water discharge (1, S_{t1}) and suspended sediment yield (2, S_{t2}) variability in the Penzhina River (Kamenskoe gauge, **a**), the Khairyuzova River (Khairyuzovo gauge, **b**), the Udova River (Rus' gauge, **c**), the Avacha River (Elizovo gauge, **d**), the Kirganick River (Kirganick gauge, **e**), the Kamchatka River (Bol'shie Sheki gauge, **f**).

reaches 25 % (except Petropavlovsky Mayak with the increase of R-factor for 4 %). The great difference of R-factor variability is observed in the Kamchatka River basin. For some meteorological stations it almost hasn't changed (the difference is less than 1 %), while for other stations it reaches 28 %.

The main factors of the sediment yield formation and variability in the Kamchatsky Krai are precipitation (amount, intensity and frequency), relief characteristics, soil and vegetable cover, and volcanic processes (Kuksina and Alexeevsky, 2018). And while such factors as relief, soil and vegetation are relatively stable in the region (expect endogenous volcanic activity with its occasional revelation), precipitation characteristics varies greatly in space and time in the territory, so precipitation, its characteristics, i.e. R-factor, is the driving factor of sediment yield formation and variability. Exclusion are territories under strong volcanic impact where sediment yield and factors of its formation and variability (e.i. relief, soils and vegetation) can change dramatically after volcanic eruptions (Kuksina and Alexeevsky, 2017).

4 Variability of suspended sediment yield

Significant trends in suspended sediment yield variability were revealed on the basis of difference-integral curves analysis. These fluctuations were compared with changes in water runoff and meteorological characteristics. Analysis of difference-integral curves for river basins of various areas demonstrates these patterns in Kamchatka rivers appears to be independent of basin area.

Analysis of difference-integral curves for rivers in the north of the Kamchatsky Krai (continental part of the region) was implemented on the basis of data of 59 years duration for the Penzhina River (Kamenskoe gauge, basin area is 71 600 km^2). Duration of observations for other rivers in the area doesn't exceed 18 years, and the revelation of long-term trends is difficult for river runoff characteristics. The Penzhina River is characterized by two relatively long trends in suspended sediment yield trends. There was an increase in SSY between 1957 and 1974, with a subsequent decline between 1975 and 2012 (Fig. 5a). These trends are very well coincided with fluctuations of water runoff and precipitation

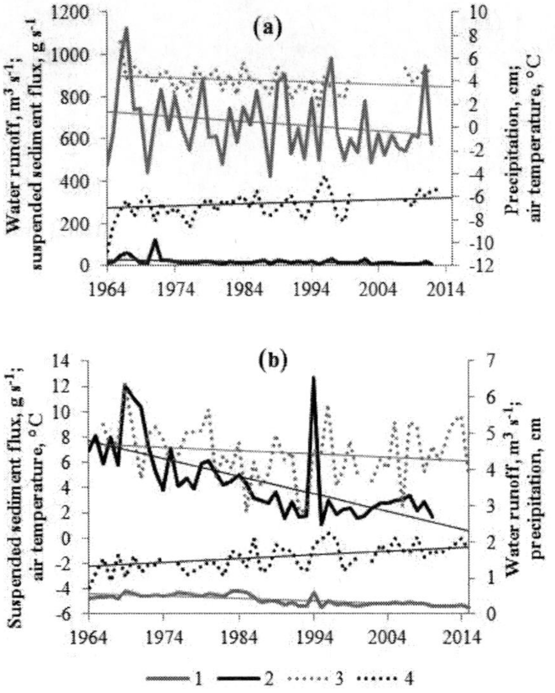

Figure 6. Correspondence of water runoff (1), suspended sediment yield (2), precipitation (3) and air temperature (4) variability in the Penzhina River (Kamenskoe, **a**) and the Dolinovka River (Dolinovka, **b**).

amount. There are three relatively long trends in water runoff variability; water runoff increased from 1957 to 1969, appeared close to be normal between 1970 and 1997, and declined from 1998 to 2010. These trends are very well timed with precipitation variability. Analysis of the data after 2010 revealed new increase in water runoff, and growth of precipitation at the same period. Reduction of water runoff in average reached 15 %, sediment yield – 41 %.

Rivers in the western part of the peninsula are characterized by synchronous variation in long-term trends of water runoff and SSY. The maximum duration of observations (57 years) is in the Khairuzova River basin (Khairuzovo gauge, basin area is 6680 km^2). The trends display no long-term trends; water runoff and SSY increased from 1976 to 1981, appeared close to normal between 1982 and 1996, declined from 1997 to 2009, and increased again after 2010 (Fig. 5b). Variability of precipitation amount, water runoff and suspended sediment yield are synchronous (against the background of heightened annual precipitation variability).

Rivers in the south-western part of Kamchatka are characterized by less coincidence in water runoff and suspended sediment yield variations. River runoff increased till the late 1970 – early 1980 in the region, and then it started to decrease, and water runoff continue to decrease nowadays, while suspended sediment yield started to increase in 2010 (Fig. 5c). Reduction of water runoff in average reached 25 %, sediment yield – 32 %.

Long-term trends in water runoff and suspended sediment yield are synchronous for rivers in the eastern part of the peninsula (with shift in start and end of phases with characteristic variation in river runoff). The longest observations are in the Avacha River basin (Elizovo gauge, basin area 4750 km^2) with duration of 78 years. In the eastern part of the peninsula, water runoff and SSF increased synchronously between 1941 and 1977, and then subsequently declined (Fig. 5d) till 2009; in 2010 has started increase against the background of precipitation growth. Reduction of water runoff in average reached 6 %, sediment yield – 49 %

The maximum amount of gauges (21) with representative monitoring data on water runoff and suspended sediment yield with duration of observations from 37 to 78 years is in the central Kamchatka region. In this part of the peninsula two phases of water runoff and SSY fluctuations occurred; increased runoff was observed till the late 1970s, followed by a period of reduced runoff (till 2009); the tendency changed to increased runoff again in 2010 (Fig. 5e). Variability of precipitation amount, water runoff and suspended sediment yield are synchronous (against the background of heightened annual precipitation variability). Reduction of water runoff in average reached 2, sediment yield – 21 %. This regularity is disturbed in rivers under impact of volcanic eruptions. For these rivers two phases of water runoff and SSF fluctuations occurred; reduced runoff was observed from the mid 1950s to the 1960s, followed by a period of increased runoff (Fig. 5f). Such response is mainly connected with the major eruption in Kamchatka in XX century of Bezymyanny volcano in March, 1956 (Gorshkov and Bogoyavleskaya, 1965). Rivers draining flanks of active volcanoes are characterized by quite synchronous fluctuations of water runoff and sediment yield.

Air temperature variability is in antiphase with water runoff and suspended sediment yield characteristics fluctuations (Fig. 6a, b).

5 Conclusions

On the basis of data on air temperature, precipitation, water runoff and suspended sediment yield analysis the main trends in sediment yield variability were revealed for the Kamchatsky Krai territory. Precipitation influences on sediment yield variability significantly through rainfall erosive factor (direct influence on sediment yield through sediment washout from river basins) and precipitation amount (indirect influence on sediment yield through water runoff), while variation of air temperature is in antiphase with water runoff and sediment yield characteristics. The decrease of precipitation is observed all over the Kamchatsky Krai territory (expect one station in the eastern coast with positive anomaly of precipitation), and it's very well coincided with fluctuations of water runoff and sediment yield in the territory. Most part of the rivers in the region is characterized by two relatively long-term trends in water runoff and sediment yield fluctuations. Increase of these characteristics was observed till the late 1970 – early 1980, and than decrease of them was registered till 2010. Analysis of the data after 2010 (till 2015) revealed new increasing trend in water runoff and sediment yield which is confirmed with precipitation variability. According to decrease of precipitation amount decrease of R-factor is observed in the Kamchatsky Krai (Kuksina and Alexeevsky, 2016). The value of this reduction varies from less than 1 % (in some station in the Kamchatka River basin) to 45 % (in the continental part of the region). Obviously decrease of R-factor leads to decrease of erosion rates and sediment yield consequently due to R-factor is driving force and the most changeable factor of sediment yield formation and variability in the Kamchatsky Krai. Exclusion is territories under strong impact of volcanic processes where volcanic eruptions can dramatically influence on sediment yield characteristics and factors of its formation.

Data availability. Underlying research data on air temperature and precipitation in available at http://meteo.ru/ (last access: 1 July 2019). Data on water discharge and suspended sediment yield is not opened for free access.

Author contributions. Author collected and processed data on air temperature, precipitation, water discharge and suspended sediment yield in Kamchatka and revealed main regularities in those characteristics variability in the region.

Competing interests. The authors declare that they have no conflict of interest.

Special issue statement. This article is part of the special issue "Land use and climate change impacts on erosion and sediment transport". It is a result of the ICCE Symposium 2018 – Climate Change Impacts on Sediment Dynamics: Measurement, Modelling and Management, Moscow, Russia, 27–31 August 2018.

Acknowledgements. I'm grateful to Nikolay I. Alexeevsky (deceased), my supervisor, with whom we started that research of Kamchatsky Krai rivers and sediments.

Financial support. This research has been supported by the Russian Scientific Foundation (project no. 17-77-10047).

References

Carrivick, J. L., Manville, V., Graettinger, A., and Cronin, S. J.: Coupled fluid-dynamicssediment transport modelling of a Crater Lake break-out lahar: Mt. Ruapehu, New Zealand, J. Hydrol., 388, 399–413, 2010.

Evstigneev, V. M.: River runoff and hydrological estimations, MSU Press, Moscow, Russia, 1990 (in Russian).

Gorshkov, G. S. and Bogoyavlenskaya, G. E.: Bezymyanny volcano and features of its last eruption in 1955–1963, Moscow, Russia, 1965 (in Russian).

Gran, K. B. and Montgomery, D. R.: Spatial and temporal patterns in fluvial recovery following volcanic eruptions: channel response to basin-wide sediment loading at Mount Pinatubo, Philippines, GSA Bulletin, 117, 195–211, 2005.

Groisman, P. Y., Knight, R. W., Easterling, D. R., Karl, T. R., Hegerl, G. C., and Razuvaev, V. N.: Trends in intense precipitation in the climate record, J. Clim., 18, 1326–1350, 2005.

Helsel, D. R. and Hirsch, R. M.: Statistical methods in water resources. Techniques of Water-Resources Investigations of the United States Geological Survey, Book 4, Hydrologic Analysis and Interpretation, chap. A3, 522 pp., 2002.

Huntington, T. G.: Evidence for intensification of the global water cycle: review and synthesis, J. Hydrol., 319, 83–95, 2006.

Korup, O.: Earth's portfolio of extreme sediment transport events, Earth Sci. Rev., 112, 115–125, 2012.

Kraevaya, T. S.: Short characteristic of mudflow regions in Kamchatka, KazNIGMI Proceedings, 33, 144–151, 1969 (in Russian).

Kuksina, L. V. and Alekseevski, N. I.: Space and Time Variations of the Runoff of Kamchatka Krai Rivers, Water Resour., 43, 438–447, 2016.

Kuksina, L. V. and Alekseevski, N. I.: The transformation of the runoff of the suspended load in Kamchatkan rivers due to volcanism, J. Volcanol. Seismol., 11, 59–70, 2017.

Kuksina, L. V. and Alexeevsky, N. I.: Spatial and temporal variability of suspended sediment yield in the Kamchatka Krai, Russian Federation, IAHS-AISH P., 367, 304–311, 2014.

Kuksina, L. V. and Alexeevsky, N. I.: Suspended sediment yield transportation by rivers of the Kamchatsky Krai into the Pacific Ocean, the Sea of Okhotsk and the Bering Sea, International Journal of River Basin Management, 17, 37–47, https://doi.org/10.1080/15715124.2018.1446963, 2018.

Kuksina, L. V., Golosov, V. N., and Kuznetsova, Yu. S.: Cloudburst floods in mountains: state of knowledge, occurrence, factors of formation, Geogr. Nat. Resour., 38, 20–29, 2017.

Larionov, G.A.: Erosion and deflation of soils: major regularities and quantitative estimation, Moscow, Russia, 200 pp., 1993.

Lavigne, F. and Thouret, J. C.: Sediment transportation and deposition by rainfall triggered lahars at Merapi Volcano, Central Java, Indonesia, Geomorphology, 49, 45–69, 2002.

Leonov, A., Leonov, V., and Hobart, J.: Valley of Geysers, Kronotsky Reserve, Kamchatka: features seen by the 1991 GOSA Expedition and Changed by the 2007 Landslide, 2013 Cyclone and 2014 Landslide, The Newsletter of the Geyser Observation and Study Association (GOSA), 28, 5–26, 2014.

Major, J. J.: Posteruption suspended sediment transport at Mount St. Helens: decadal-scale relationships with landscape adjustments and river discharges, J. Geophys. Res., 109, F01002, https://doi.org/10.1029/2002JF000010, 2004.

Major, J. J. and Mark, L. E.: Peak flow responses to landscape disturbances caused by the cataclysmic 1980 eruption of Mount St. Helens, Washington, GSA Bulletin, 118, 938–958, 2006.

Major, J. J., Crisafulli, C. M., Frenzen, P., and Bishop, J.: After the disaster: The hydrogeomorphic, ecological, and biological responses to the 1980 eruption of Mount St. Helens, Washington, in: Volcanoes to Vineyards: Geologic Field Trips through the Dynamic Landscape of the Pacific Northwest: Geological Society of America Field Guide, O'Connor, edited by: O'Connor, J. E., Dorsey, R. J., and Madin, I. P., 15, 1–24, 2009.

Manh, N. V., Merz, B., and Apel, H.: Sedimentation monitoring including uncertainty analysis in complex floodplains: a case study in the Mekong Delta, Hydrol. Earth Syst. Sci., 17, 3039–3057, https://doi.org/10.5194/hess-17-3039-2013, 2013.

Mueller, E. N. and Pfister, A.: Increasing occurrence of high-intensity rainstorm events relevant for the generation of soil erosion in a temperate lowland region in Central Europe, J. Hydrol., 411, 266–277, 2011.

Report on climate risks in the Russian Federation territory, Saint-Petersburg, 106 pp., 2017 (in Russian).

Shkaberda, O. A. and Vasilevskaya, L. N.: Long-term changes of temperature and humidity regime in Kamchatka Peninsula, TINRO Proc., 178, 217–233, 2014 (in Russian).

Wischmeier, W., Johnson, C., and Cross, B.: A soil erodibility nomograph for farmland and construction sites, J. Soil Water Conserv., 26, 189–193, 1971.

Wischmeier, W. H. and Smith, D. D.: Predicting rainfall erosion losses: A guide to conservation planning, Agric, Handbook, 67 pp., 1978.

Proc. IAHS, 381, 65–71, 2019
https://doi.org/10.5194/piahs-381-65-2019

Numerical modeling and forecast of channel changes on the river Lena near city Yakutsk

Pavel Golovlyov[1], Ekaterina Kornilova[1], Inna Krylenko[1,2], Vitaly Belikov[2], Aleksandr Zavadskii[1], Eugeniya Fingert[1,2], Natalya Borisova[2], and Elizaveta Morozova[1]

[1]Faculty of Geography, Lomonosov Moscow State University, Moscow, 119991, Russia
[2]Water Problems Institute, Russian Academy of Sciences, Moscow, 119333, Russia

Correspondence: Pavel Golovlyov (pavel_golovlev@list.ru)

Published: 1 August 2019

Abstract. STREAM_2D software package was applied to retrospective and predictive simulations of the Lena River near city Yakutsk hydraulic and channel changes during ice-free period. The modelling results indicate significant correspondence of simulated water discharges distribution and water levels with observed ones for the period 2001–2016. Model has captured main erosion and depositional zones observed in the 2009–2016 years. Combination of typical monthly average hydrographs of 1 %, 10 % and 50 % return intervals was used as an initial parameter for 10-year channel changes forecast. According to the simulation results, degradation of the Adamovskaya branch will be continued, which is the most negative impact for the Yakutsk city infrastructure maintenance. At the end of the forecast period the equal distribution between Adamovskaya and Buorylarskaya channels is possible, which is a positive trend from the point of view of the city water supply system.

1 Introduction

In the city of Yakutsk on the Lena River numerous engineering and water facilities exist, for the safe operation of which it is necessary to take into account the peculiarities of the water and channel regime of the river. In recent years, an unfavourable situation for the entire water infrastructure of the city has been created in the territory, due to the fact that the river channel near the left bank, where the city is located, grows shallower, while the right branch develops and large forms of the channel topography also shift (Chalov, 2016). The above-mentioned re-formations of the channel lead to the displacement of the navigable fairway, which complicates the work of the port, leads to failure of the city water intake, and complications associated with the ferry crossing. For the first time the adverse channel changess effect on this section of the Lena River was discovered as early as the 1970s by the expedition of the Lomonosov Moscow State University, and even then were made forecasts, according to which the shallowing of the branches near the city was expected (Chalov, 2012). In connection with the existing problems of water use, different projects were performed for the preserving of the main flow in the city area. Large number of papers were devoted to the study of channel processes in this area (Chalov, 2012, 2016; Zaitsev and Chalov, 1989), both with hydrological model applications (Zaitsev et al., 2004). This study is aimed at the significantly larger extent of the modelling area and the considered period for study area on the base of STREAM_2D hydrodynamic model (Belikov and Kochetkov, 2014), adapted for the 75 km long Lena River section for the period 2001–2016 years. Due to its complexity, hydrodynamic models with movable bottom are mainly used for non-extended channel sections and/or for non-extended periods (Sloff et al., 2012; Horvat et al., 2015; Zhang et al., 2014), often the data about real channel changes for long period are not available for the channel deformation modelling block validation. Therefore, to apply the model for a complex channel and floodplains system of the Lena River near Yakutsk the first task was detailed model hydraulic calibration and validation. Next goal was validation of the channel deformation modelling through continuous simulation for the period between repeated channel surveys of 2009 and 2016 in order to assess the quality of the model's reproduction of the main trends in vertical channel changes. The final goal of the research was forecast estimations for the possible trends in channel changes over the next decade, especially in view of the difficult situation with the city's water supply due to the shallowing of the Adamovskaya channel closest to Yakutsk city.

2 The study area

The city Yakutsk, located in the Lena River valley in the area under study, is an old Russian city (founded in the 17th century), is now the administrative centre of the Sakha Republic. The city's population is about 300 thousand people. Within the considered section of the Lena River near Yakutsk city, three hydrological gauges are located (Fig. 1). The measuring discharges and water levels gauge in the village of Tabaga is located in the upper part of the studied reach. The Lena River basin area for the Tabaga gauge is 897 000 km^2. The water level measuring gauge of the Yakutsk city is situated in the middle part of the study area, the Kangalassy gauge is located at the lower boundary of the area (Fig. 1). The study section of the river is characterized by numerous ramifications, and is located between the Tabaginsky cliff at upstream and the Kangalassky cliff in the lower reaches, which limit the spread of the left bank floodplain and low terraces above the floodplain, as they are rocky capes. The width of the main channel of the Lena River in this section is 3.5–4 km.

Near the city of Yakutsk, the Ponomaryov Island (Fig. 1) divides the main channel into two branches, comparable in terms of width and water flow – the Adamovskaya and Buorylarskaya branches. The water content of the Adamovskaya branch, which is closest to the city, has recently been degraded due to less water coming there (Chalov, 2012). In this section the floodplain width is 6–7 km, the floodplain height reaches 10 m above the low-water level, the floodplains are developed insignificantly and the city of Yakutsk is mainly located on unflooded elevations.

The annual water level regime of the Lena River near Yakutsk city is characterized by a pronounced rise of the water levels during the spring flood, including because of ice jams, significant and sharp rises in water levels during summer-autumn floods and a relatively low and stable level in the cold part of the year. The ice regime of the Lena River has a number of features that are caused by the severe, extremely continental climate and the spread of permafrost. The ice jams, lasting from one to 5 d emerge on average every 1 to 2 years near the Yakutsk city in the areas of channel ramifications.

The maximum water discharge was observed on 24 May 1966 (50 900 $m^3\,s^{-1}$); and in the last two decades close water discharges amounted to 49 600 $m^3\,s^{-1}$ were recorded on 11 June 2006 and 20 May 2010. In general, there is a statistically significant increase in both average annual and maximum water discharge, which becomes particularly important from the beginning of 21st century.

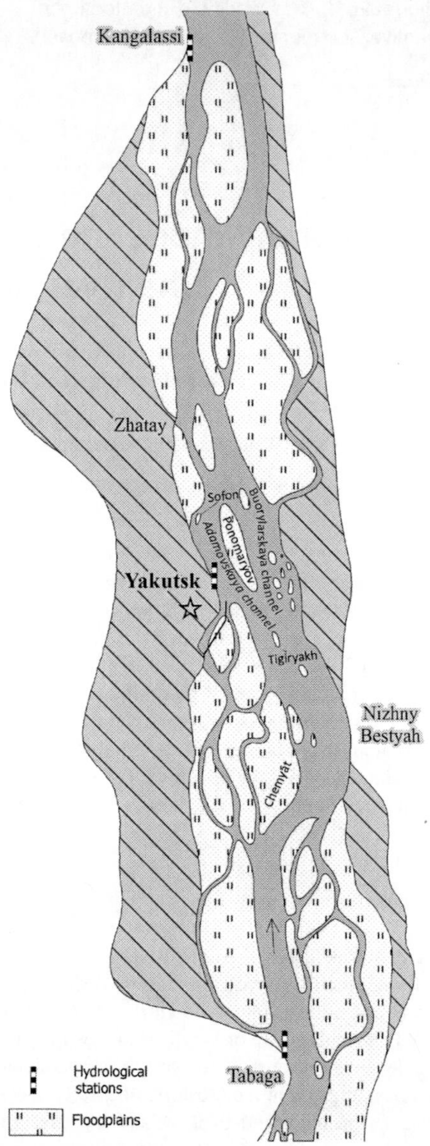

Figure 1. Study area.

3 Methods and data

STREAM_2D software package, widely used in Russia (Belikov et al., 2015), based on the numerical solution of two-dimensional Saint-Venant equations in the "shallow water" approximation using hybrid computational grids of irregular structure and original topography interpolation algorithms (Belikov and Semenov, 2000) was used for modelling. This software showed high efficiency in modelling both for flow hydraulics and channel changes, including those for different key area in Russia (Volga, Ob, Lena, Kolyma, Vilyuy, Yana, Amur etc.) (Zaitsev et al., 2004; Belikov et al., 2002; Krylenko et al., 2018).

The description of the mathematical equations included in the numerical scheme of the model is presented, for example, in Belikov et al. (2015) and Aleksyuk and Belikov (2017). When modelling, a block of channel changes was involved, taking into account the convective transport of soil particles by flow, soaking up and sedimentation of sediments in an uneven flow, changing the bottom levels over time, taking into account the effect of flattening (transverse diffusion) of the underwater slope in a direction orthogonal to the velocity vector.

The initial data for the application of the hydrodynamic model were the materials of field researches of the soil erosion and channel processes laboratory at the Geography Faculty of Moscow State University, including the results of bathymetric surveys, measurements of water discharges, water levels and water slopes; daily regime observation data of hydrological gauges on the Lena rivers – Tabaga, Yakutsk, Kangalassi. To create the integrated digital topography models, the bathymetric survey data for 2009 and 2016 were integrated with the digital model of the floodplain topography, obtained in 2016 (based on topographic maps data and field surveys of the topography).

The map of jam-hazardous part of river channels was compiled on the basis of historical data and satellite images to set the places of ice jams formation, the dates of the ice jams formation were taken according to the hydrological gauges data.

The field data obtained during the expedition researches in 2016 was used to set the parameters of the channel-forming sediments fractions. Mainly the channel sediments are represented by the sand with an average diameter of 0.36 mm, so the average diameter of the channel sediments was taken into account in the simulations.

At this stage, the modelling was carried out only for the warm high water period from 1 May to 31 October of each year, which covered the passage of the most important for the formation and displacement of the main channel forms of water flow floods and high waters.

4 Results of calibration and validation of hydraulic characteristics

The calibration and verification of the model was originally carried out based on data from 2016 field surveys, including a comparison of the water level marks along the longitudinal profiles and the distribution of the water discharges along the branches for various water discharges. When the roughness coefficients equal to 0.015–0.017 for the channels and 0.05 for the floodplain, a good compliance was reached between the measured and calculated water levels at the control points. The difference between the modelled and observed water levels at the Yakutsk gauge was 1–4 cm. On the average along the length of the study section, the difference between the values obtained during special expeditionary surveys (instantaneous water level measurements) and the results of modelling at control points did not exceed 10–18 cm.

The model reproduces well the distribution of water discharges among the main branches (Table 1). In the main branching point near the Yakutsk city, which determines the water situation of the city, the Adamovskaya channel, closest to the city, receives 37 %–39 % of the input water discharge in the flow range from 9560 to 30 000 $m^3 s^{-1}$ according to measurement data, the right Buorylarskaya channel gets up to 55 % of the input flow, which reproduces the model. The remaining 7 % of the input water flow goes along the Khatykhstakhaya channel, and the model slightly underestimates the water discharge in it (modelling value is 5 % of the total water flow), which is due to the small detailing of the computational grid in this channel due to its insignificant width in comparison with the main channel.

Additional refinement of the roughness coefficients for the ice jams period at the hazardous sections according to the data for 2009–2015 was made to simulate water flow dynamics for the flood periods, and it was found that the best compliance of the water levels is provided with the roughness coefficients of the ice jams sections equal to 0.073.

The simulation, based on data for the period from 2001–2015, showed that the hydrodynamic model, adapted according to the field and regime observations data for 2009–2016, provides a steady good quality of water levels modelling over a long-term period, which confirms both by visual match of the graphs of the observed and simulated water levels (Fig. 2), and high values of the simulation quality criterion. The criterion of Nash-Sutcliffe (NSE) for the studied periods for the Lena River gauge in Yakutsk was 0.90 for 2001–2008 and 0.93 for 2009–2015 years (in the hydrological modelling practice it is assumed that the results with NSE > 0.75 can be considered as good).

An additional validation of the model was carried out by comparing the simulated flooding boundaries with the flooding boundaries provided by satellite images (more than 20 images). It was found that the relative error when comparing the real hydrological situations observed by modeling and interpreting satellite images the divergence of flooded areas does not exceed 10 % during floods and high waters and the maximum divergence reaches 18 % during ice jams (Kornilova et al., 2018).

5 Validation of the channel changes modelling results for the historical period

Due to the presence of multi-temporal bathymetric surveys, both actual and model bottom evolution were evaluated for the period 2009–2016. Despite some limitations of model simulations (only the warm period of the years were mod-

Table 1. Flow distribution according observation and modeling.

No.	Date	1 Jun 2016				2 Jul 2016				19 Aug 2016			
	Section	Q_{obs}		Q_{sim}		Q_{obs}		Q_{sim}		Q_{obs}		Q_{sim}	
		$m^3 s^{-1}$	%	$m^3 s^{-1}$	%	$m^3 s^{-1}$	%	$m^3 s^{-1}$	%	$m^3 s^{-1}$	%	$m^3 s^{-1}$	%
1	Input discharge	30 000	100	30 000	100	18 840	100	18 840	100	9560	100	9560	100
2	Main channel near isl. Elovy	19 800	66	20 248	67.5	–	–	15 134	80.3	–	–	9041	94.6
3	Buorylarskaya channel	16 500	55	16 564	55.2	9951	53	9936	52.7	5229	55	5322	55.7
4	Adamovskaya channel	11 100	37	11 549	38.5	7305	39	7537	40.0	3604	38	3800	39.7
5	Khatykhstakhaya channel	2400	8	1545	5.2	1357	7	960	5.1	686	7	400	4.2
6	Right channel near isl. Sakhalin	7500	25	8543	28.5	4241	23	5392	28.6	2897	30	2924	30.6
7	Zhatay channel	7200	24	5500	18.3	2397	13	2307	12.2	629	7	575	6.0
8	Main channel near isl. Kopilka	24 300	81	25 272	84.2	–	–	16 883	89.6	–	–	9340	97.7

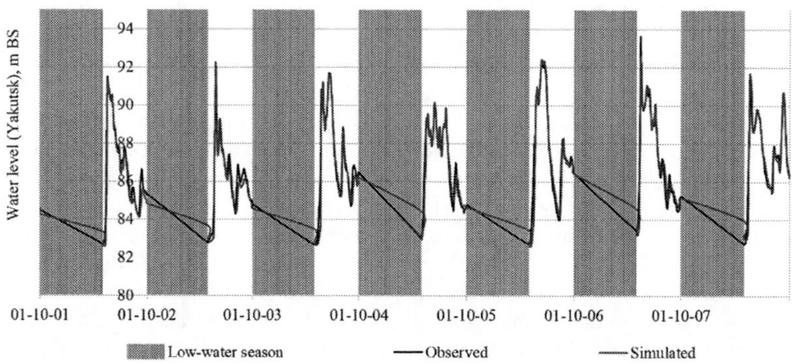

Figure 2. Observed and modelled water levels at the Yakutsk gauge at 2001–2008 (ice-free periods).

elled, sediments were considered as homogenous), the main positive result was the correspondence of the simulated and observed main zones of erosion and deposition (Fig. 3). Totally, in the 70 % of modeling sells inside the channels deposition or erosion zones coincide with estimated according bathymetry surveys. But one can notice some alternation of sections with almost complete coincidence of the situation of the simulated and actual changes and areas, where the model reproduced not so good (Fig. 3).

The model adequate showed the channel changes in the upstream part of the studied section from the Tabaga until the island Chemyat. Downstream from this part of the channel model indicated erosion zone near the right steep bank, but in reality there are deposition zone. The possible reason of this discrepancy can be that the model is underestimated planform channel changes, which took more flow energy. There is the vast deposition zone near the Tigiryakh island. The stable two-stream system has formed near this island in the last decade, which is of decisive importance for the channel processes within the Yakutsk. In this area, the model again showed a situation close to reality. The obtained distribution of water flow along the branches corresponds to that observed in the 2016 year (the flow discharge was distributed equally among the right and left main branches). Unfortunately, in the next section from the island Tigiryakh to Ponomarev island, which is the most difficult for modelling because of the diversity channel processes factors, the simulation results can only be called partially successful. The model showed erosion along the left bank, but did not identify it in the central part of the channel. Despite the strong discrepancy between the actual and model data, the distribution of water flow between Adamovskaya and Buorylarskaya branches coincides with the observed (39 % and 51 % correspondingly). Along the Adamovskaya and Buorylarskaya channels, the model has demonstrated quite realistic results. In the downstream section of the Lena River near Zhatay

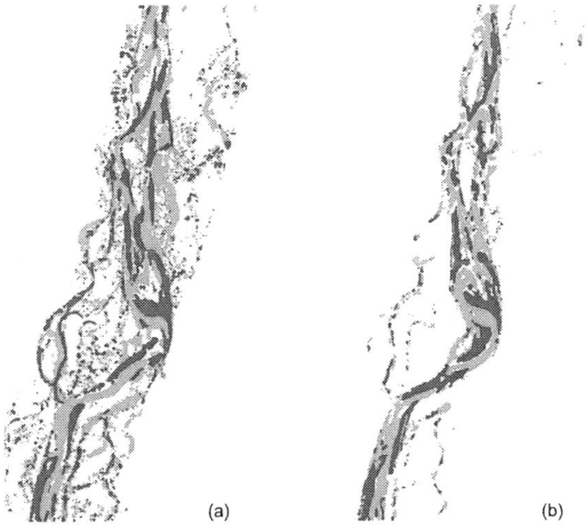

Figure 3. Deposition (shown in green colour) and erosion (shown in red colour) zones in the Lena River channel for the period 2009–2016 year according observation data **(a)** and modelling **(b)**.

75 % of the flow is going in the main stream. Modelling on this section showed a result that is different from the observed one. This section is characterized by the maximum rate of planform changes. Thus, we can conclude about the connection of "unsuccessful" simulation results in some sections with the determining role of planform changes in these sections. The further development of the model should take into account planform channel deformation block improvement, including taking into account transverse circulation in water flow. Some discrepancies between model results and real changes can be connected with neglecting of the presence of permafrost, which is very difficult to take into account in such kind of modelling.

6 Channel evolution forecast

Ten-year prediction of channel changes of the Lena River near Yakutsk city considered the combination of typical hydrographs of different probabilities (50 %, 10 %, 1 %). The input hydrographs for each year was built on the base of monthly averages discharges (with the exception of 6 winter months from November to April). Then from these annual stepped hydrographs a combined hydrograph was compiled for a ten-year period in the following sequence: 50 %, 10 %, 50 %, 1 %, 50 %, 50 %, 50 %, 10 %, 50 %, 50 %. Further, this stepped hydrograph was approximated by a piecewise linear continuous equivalent hydrograph, which gave the averaged values of water discharges for each month coinciding with the monthly average discharge. The topography of the channel of 2016 was used as input in the beginning of the forecast period.

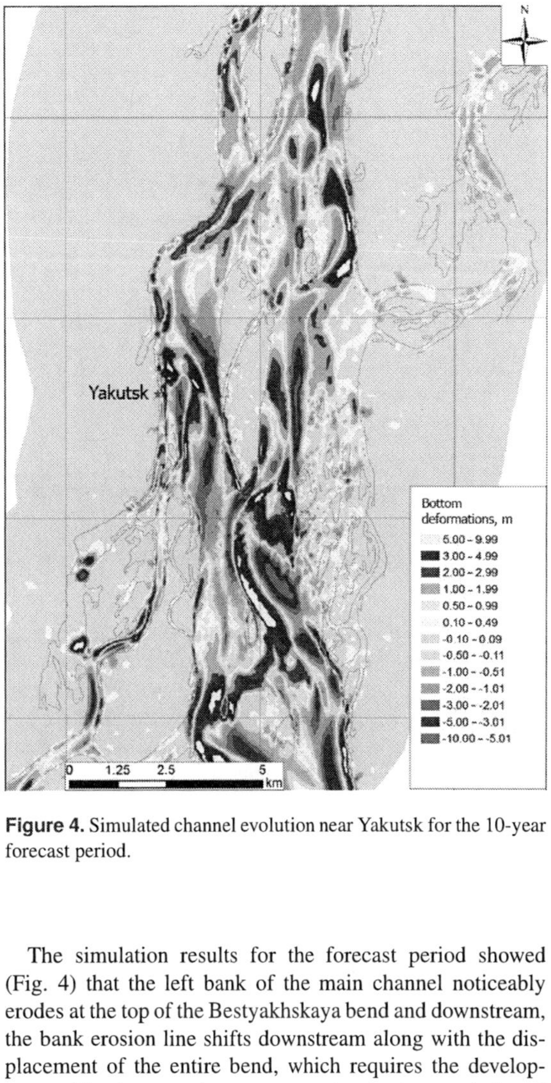

Figure 4. Simulated channel evolution near Yakutsk for the 10-year forecast period.

The simulation results for the forecast period showed (Fig. 4) that the left bank of the main channel noticeably erodes at the top of the Bestyakhskaya bend and downstream, the bank erosion line shifts downstream along with the displacement of the entire bend, which requires the development of bank protection measures due to the proximity of the highway. Previously, this section of the bank was already protected by spurs, but at present they are not effective due to the channel shift. At the same time, the right channel which is used for shipping presently below Nizhny Bestyah Island is activated and slightly straightened according to the simulation results, which allows to predict that in the coming years it will remain the most deep and navigable. The channel that straightens the bend in the area of the Chemyat Island is actively developing in the entrance, but then falls into the sediment deposition zone.

According to the simulation results for the 10-year forecast period, the Adamovskaya channel near Yakutsk will be degraded. This is have a harmfull effect to the maintenance of the city water intake system, due to the sandbar formation near the left bank and its downstream displacement (by ap-

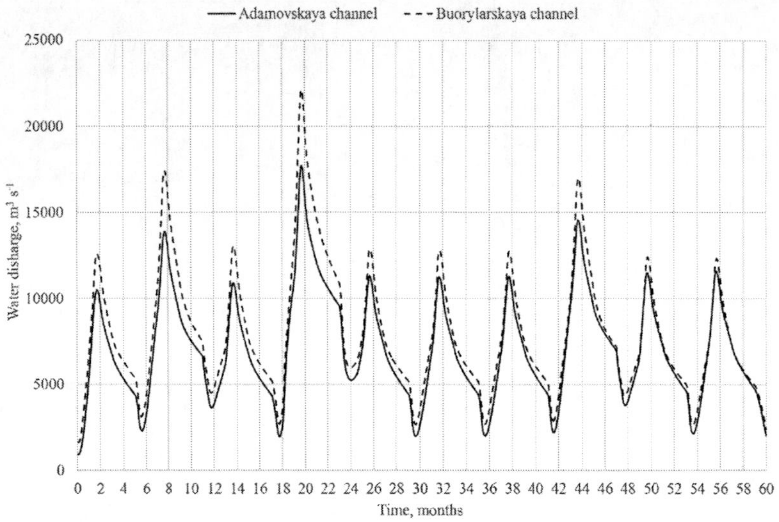

Figure 5. Distribution of water discharges between Adamovskaya and Buorylarskaya channels for 10-year forecast period.

proximately 500–800 m for 10 years). This will lead to huge deposition at the Yakutsk river harbour, and blocking of the clean water to the water intake from the main channel in the winter low-water period. At the same time, the intensive erosion of the bottom on the right side of the Adamovskaya channel (near the Ponomarev Island) and the displacement of the line of maximum depths from the left bank of the Adamovskaya channel (i.e. from the Yakutsk city) to the island will continue.

The upper part of the Ponomarev Island is intensively eroded, and downstream the Sofon Island is intensively eroded too (Fig. 4). The forecasts show that the last one will disappear due to erosion. If we analyse the diagram of water discharges distribution at the bifurcation nod of the Ponomarev Island (Fig. 5) over the ten-year forecast period, it is clear that at present about 10 %–15 % less water goes to the Adamovskaya channel than to the Buorylararskaya. However, after 10 years the water discharges are almost equal. This is due to an increase of a massive side bank in the right channel, which begins to block the entry into the Buorylarskaya channel, which is clearly seen in Fig. 5. This tendency will create some advantages perspective in the city water supply.

Intensive erosion is predicted in the area of the Zhatay village, approximately 2 km below the village, on the right bank of the main channel. At the same time, the process of alluviating of the access channel to the boatyard of the Zhataisky ship-repairing plant will continue, which requires conducting of the channel-improvement measures.

7 Conclusions

The numerical two-dimensional hydrodynamic model of the Lena River near Yakutsk city was developed, calibrated, and verified using field data. The model successfully reproduce the distribution water discharges among the channels, the water level regime for a long-term period and zones of inundation.

Comparison of modelled bottom changes with the observed ones for the period 2009–2015 has demonstrated that in the 70 % of channel grid sells model adequately reproduce the erosion and deposition trends. In connection with the novelty of such studies for extended sections of large sections of lowland rivers, such results should be considered as positive.

Predictive simulations of the hydraulic flow regimes and the bottom evolution in the ice-free period were carried out on the basis of typical monthly average hydrographs of 1 %, 10 % and 50 % return intervals. Based on the simulation results, Adamovskaya channel will be further degraded, but to the end of 10 years forecast period the equalization of water discharges entering the Adamovskaya and Buorylarskaya channels is possible.

The channel development forecasts obtained as a result of the simulation served as the basis for the development of the hydrotechnical measures patterns to reduce the adverse effects of the channel processes for Yakutsk area.

The low-water (mostly subglacial) channel changes cover a much smaller part of the river channels and do not determine the main trends in the development of the channel. At the same time, the modelling of channel changes taking into account the ice cover as well as permafrost are separate, very complex and important tasks that are planned to be solved at the next stages of the studies, which are beyond the scope of this publication.

Data availability. Data available upon request to the contact author.

Author contributions. VB developed mathematical model, AZ, VB, IK developed experiment design, PG, EM prepared digital terrain models, PG, KE, IK, EF performed the simulations and result analysis, NB performed forecast simulations, AZ, EM analyzed historical channel changes.

Competing interests. The authors declare that they have no conflict of interest.

Special issue statement. This article is part of the special issue "Land use and climate change impacts on erosion and sediment transport". It is a result of the ICCE Symposium 2018 – Climate Change Impacts on Sediment Dynamics: Measurement, Modelling and Management, Moscow, Russia, 27–31 August 2018.

Acknowledgements. The authors of the article express their deep gratitude to the Laboratory of Soil Erosion and Channel Processes of the geographical faculty of Moscow State University for providing the field research materials.

Financial support. Simulation of the flow dynamics was supported by the Russian Foundation for Basic Research (grant RFBR no. 17-05-01230), the channel changes evaluation was supported by the Russian Science Foundation (grant RSF no. 18-17-00086).

References

Aleksyuk, A. I. and Belikov, V. V.: Simulation of shallow water flow with areas and bottom discontinuities, Comp. Math. Math. Phys., 57, 318–339, https://doi.org/10.1134/s0965542517020026, 2017.

Belikov, V. V. and Kochetkov, V. V.: Software Complex STREAM_2D to Calculate Streams, Bottom Deformation, and Pollutants Transfer in Open Flows, Software State Registration Certificate no. 612181, Russian Agency for Intellectual Property, 2014.

Belikov, V. V. and Semenov, A. Yu.: Non-Sibsonian interpolation of arbitrary system of points in Euclidean space and adaptive isolines generation, Appl. Numer. Math., 32, 371–387, 2000.

Belikov, V. V., Zaitsev, A. A., and Militeev, A. N.: Mathematical modeling of complex sections of large rivers channels, Water Resour., 29, 698–705, 2002 (in Russian).

Belikov, V. V., Krylenko, I. N., Alabyan, A. M., Sazonov, A. A., and Glotko, A. V.: Two-dimensional hydrodynamic flood modelling for populated valley areas of Russian rivers, Proc. IAHS, 370, 69–74, https://doi.org/10.5194/piahs-370-69-2015, 2015.

Chalov, R. S., Kirik, O. M., Ilyasov, A. K., and Botavin D. V.: Yakutsk water junction on the Lena river: problems of channel processes, history, modern times, prospects, management possibilities, Water Industry of Russia: problems, technologies, managemen, 3, 44–56, 2012 (in Russian).

Chalov, R. S., Zavadsky, A. S., Ruleva, S. N., Kirik, O. M., Prokop'ev, V. P., Androsov, I. M., and Sakharov, A. I.: Morphology, deformations, temporary changes in the channel of the Lena river and their influence on the economic infrastructure in the Yakutsk region, Geomorphology, 3, 22–35, 2016 (in Russian).

Horvat, Z, Horvat, M., and Spasojevic, M.: Two dimensional river flow and sediment transport model, Environ. Fluid Mech., 15, 595–625, https://doi.org/10.1007/s10652-014-9375-y, 2015.

Kornilova, E. D., Krylenko, I. N., Golovlyov, P. P., Sazonov, A. A., and Nikitsky, A. N.: Verification of the two-dimensional hydrodynamic model of the Lena River near Yakutsk according to time-based space imagery data, Modern problems of remote sensing of the Earth from space, 15, 169–178, https://doi.org/10.21046/2070-7401-2018-15-5-169-178, 2018 (in Russian).

Krylenko, I., Belikov, V., Fingert, E., Golovlyov, P., Glotko, A., Zavadskii, A., Samokhin, M., and Borovkov, S.: Analysis of the impact of hydrotechnical construction on the Amur river near Blagoveshchensk and Heihe cities using a two-dimensional hydrodynamic model, Water Resour., 45, 112–121, https://doi.org/10.1134/S0097807818050378, 2018.

Sloff, K. and Mosselman, E.: Bifurcation modelling in a meandering gravel-sand bed river, Earth Surf. Proc. Land., 37, 1556–1566, https://doi.org/10.1002/esp.3305, 2012.

Zaitsev, A. A. and Chalov, R. S.: Channel processes and regulation of the river Lena in the area of Yakutsk, Water Resour., 5, 75–81 (in Russian), 1989.

Zaitsev, A. A., Belikov, V. V., and Militeev, A. N.: Using computer modeling for a large river, Proc. Int. Symp. "Sediment Transfer through the Fluvial System", Moscow, 2–6 August 2004, IAHS Publ. 288, ISBN 1-901502-67-8, 386–394, 2004.

Zhang, W., Xu, Y., Wang, Y., and Peng, H.: Modeling Sediment Transport and River Bed Evolution in River System, Journal of Clean Energy Technologies, 2, 175–179, 2014.

Proc. IAHS, 381, 73–77, 2019
https://doi.org/10.5194/piahs-381-73-2019

Open Access
PIAHS

Two-dimensional hydrodynamic modelling of channel processes and floods characteristics at the confluence of the Amur and Zeya rivers

Eugeniya Fingert[1,2], **Inna Krylenko**[1,2], **Vitaly Belikov**[2], **Pavel Golovlyov**[1], **Aleksandr Zavadskii**[1], and **Mikhail Samokhin**[1]

[1]Faculty of Geography, Lomonosov Moscow State University, Moscow, 119991, Russia
[2]Water Problems Institute, Russian Academy of Sciences, Moscow, 119333, Russia

Correspondence: Eugeniya Fingert (fingerte@gmail.com)

Published: 1 August 2019

Abstract. Two-dimensional hydrodynamic STREAM_2D model have been applied to the confluence of Amur and Zeya rivers. Main flow characteristics and possible channel changes are considered during modeling for two scenarios – without dams near Big Heihe Island (1) and with four dams in the side channels around the island (2). Results of modeling demonstrate impact of dams construction on flow concentrating in the main channel, which can lead to increasing of flow velocities and subsequent erosion in main Amur River channel downstream the river confluence.

1 Introduction

The confluence of the Amur and Zeya rivers is densely populated area. There are two main cities – Russian city Blagoveshchensk with 200 000 residences on left bank of the river Amur and Chinese city Heihe with population over 1.5 billion residences on the right bank of the river. This territory is characterized by significant water problems related to floods and intensive channel processes. Last significant flood is affected Chinese and Russian citizens in the 2013 year, when the most part of floodplains, including many small settlements and the island Big Heihe were inundated (Danilov-Danilyan et al., 2014; Bolgov et al., 2015). According to international legislations, the border between Russia and China is marked along channel thalweg, which position is continuously changing due to natural and anthropogenic drivers. In last 10 years, this area has attracted attention due to intensification of the hydrotechnical construction at the river confluence, including embankment of Blagoveshchensk and dams around the Big Heihe Island.

Modern studies for key area are based on the methods of hydrodynamic modeling. The influence of the Blagoveshchensk embankment on the channel processes have been investigated using two-dimensional model in 2011–2012 years (Belikov et al., 2012). The Chinese Institute of Water Resources and Hydropower Engineering (IWHR) has also developed models for this area. The further investigations of channel processes using two-dimensional hydrodynamic model were done based on novel bathymetric and floodplain topography datasets and also accounts dam allocation along channels around island Big Heihe (Krylenko et al., 2018). Both mentioned studies are identified, that embankment of Blagoveshchensk is affected on the flow only locally in the section of the Amur River above the confluence. The partial cutting of the secondary channels by the dams from the right bank of the Amur River can lead to the significant partitioning of water flow and increasing of erosion in the Amur River upstream and downstream from the confluence with the Zeya River.

Following with past experiments with two-dimensional hydrodynamic models (Belikov et al., 2012; Krylenko et al., 2018) hydrodynamic model 5-year forecast of possible channel changes were performed and their influence on the evolution floods characteristics were estimated for two possible scenarios of river development: (1) without dams along channels around island Big Heihe and (2) under impacts of dams.

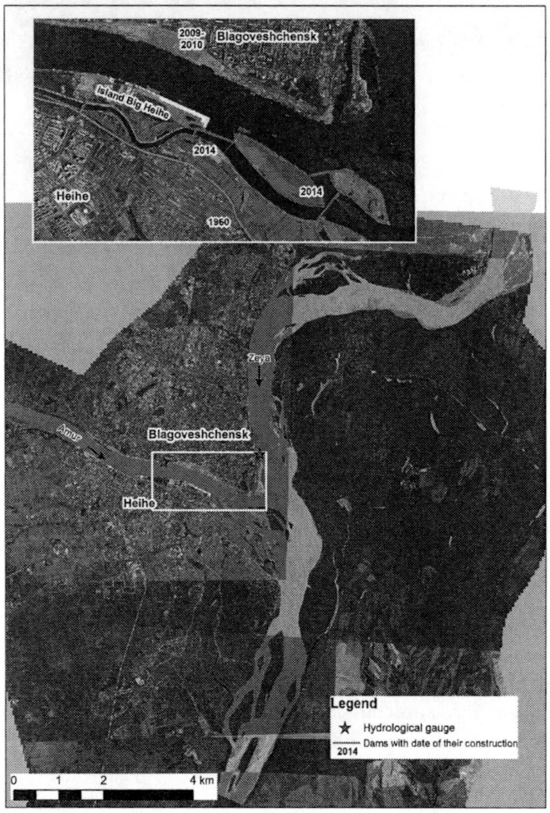

Figure 1. Study area.

2 The study area

The modeling area is included 15 km of the Amur River and 20 km of the Zeya River upstream from their confluence and 15 km of the Amur River downstream from the confluence (Fig. 1). The typical features of the Amur River channel are changing along its length. There is incised channel with many hard rock spots on the Amur River above the river confluence. Within Blagoveshchensk city Amur River channel is rectilinear, adapted along the left terrace bank, somewhere it is complicated by unilateral right-bank branches. Most of the urban area of Blagoveshchensk is located on the terrace above the floodplain. The opposite right bank, where the city Heihe is located, is characterized by the floodplains with the width about 5–6 km. The confluence of the Amur with the Zeya river have the complicated hydraulic conditions with backwater phenomena. Amur River has formed an inland delta with several islands near the right bank there. The Zeya River above the confluence as well as Amur River below the confluence is characterized by the wide floodplains with free conditions of the channel evolution.

The floods on these rivers are induced by the rains. The maximum water discharges of the Amur and the Zeya rivers at their confluence were observed in the August 1984, and were 16 700 and 12 800 $m^3 s^{-1}$, respectively. Recent extreme flood took place in the August 2013, maximum water discharges were 12 288 and 13 400 $m^3 s^{-1}$ correspondingly.

Both Blagoveshchensk and Heihe cities are protected from floods by the system of dams, constructed in the middle of 20th century. The embankment of Blagoveshchensk, which also has flood protective function as well as architecture value, is under construction from 2009–2010 years, and one of its section is not finished yet. The systems of dams around the Big Heihe Island is included four ground dams, blocking the flow in the small channels between islands from the Chinese riverside (Fig. 1). These constructions are look like ground cofferdams without any concrete reinforcements, the height of the dam's crest is the same, as the height of nearby floodplains (about 124.5 m a.s.l.), which means, that during high flow periods dams can be overflowed.

3 Methods and data

STREAM_2D software package, widely used in Russia (Belikov and Kochetkov, 2014), based on the numerical solution of two-dimensional Saint-Venant equations in the "shallow water" approximation using hybrid computational grids of irregular structure and original topography interpolation algorithms (Belikov and Semenov, 2000) was used for modeling. This model showed high efficiency in modeling both for flow hydraulics and channel deformations, including those for the rivers of the eastern region of Russia (Lena, Kolyma, Vilyuy, Yana, Amur) (Zaitsev et al., 2004; Belikov et al., 2002; Krylenko et al., 2018).

When modeling, a block of channel deformations was involved, taking into account the convective transport of soil particles by flow, soaking up and sedimentation of sediments in an uneven flow, changing the bottom levels over time, taking into account the effect of flattening (transverse diffusion) of the underwater slope in a direction orthogonal to the velocity vector. The description of the mathematical equations included in the numerical scheme of the model is presented, for example, in Belikov et al. (2015) and Aleksyuk and Belikov (2017).

The discretization of the modelling area in STREAM_2D was performed using an irregular hybrid computational mesh with more than 72 thousand cells. Curvilinear quadrangular grid was used for line structures; curvilinear triangular grid with a spatial resolution from 20×30 to 150×200 m was adopted for the channels and floodplains.

Topography of floodplains, based on the highly detailed maps and bathymetry of river channels were used as the model input. The configuration of linear infrastructure including the dams from the Chinese side, the embankments of Blagoveshchensk, existing bridges and those under con-

Table 1. Comparison of water levels for validation scenarios.

Dates	Input discharges of the rivers ($m^3 s^{-1}$)		Water levels (m a.s.l.)					
			Observed	Simulated	Difference between simulated and observed level, cm	Observed	Simulated	Difference between modeled and observed level, cm
	Amur	Zeya	Blagoveshchensk – Amur			Blagoveshchensk – Zeya		
13 June 2018 (low flow)	1510	1390	120.63	120.67	4	120.15	120.21	6
18 July 2018 (rain flood)	3500	7680	124.70	124.79	9	123.90	123.91	1
16 August 2013 (extreme rain flood)	12 288	13 400	128.10	128.21	11	127.93	127.92	−1
15 August 1984 (extreme rain flood)	16 700	12 800	128.45	128.61	16	128.29	128.24	−5

struction, and the road embankments on the floodplain were taken into account in the model grid and topography. Several digital terrain models with bathymetry 2017 and 2018 years and different sets of engineering structures were prepared. New data was also supplemented with the results of 2011 field survey.

Data from hydrological statations Zeya–Belogorie, Amur–Kumara, and Amur–Grodekovo were used as the boundary conditions. The water levels at stations, located in the cities (Blagoveshchensk on the Amur and on the Zeya), were taken into account for the model calibration and validation. The data of engineering survey of 2017–2018, including the measured water discharges, and water surface slopes, were used for model development.

The average diameter of channel sediments 0.9 mm was taken in to account, the diameter of the 90 %-probability (90 % of particles is less than 2.5 mm) was set at 2.5 mm for the modeling. The position of the surface of clay sediments and rocky cliffs was defined according the geological survey data.

The simulation was carried out for a 5-year forecast period, daily hydrographs of the previous period 2011–2015 years were used as the boundary conditions. This period can be assumed as the most representative, because it is included the extreme flood period of the year 2013 as well as years with the average flow conditions.

4 Model calibration and validation

The initial model calibration and validation were performed using bathymetry data from the 2017 year and demonstrated good correspondence of modelled and observed water levels for different hydrological situations (Krylenko et al., 2018). On this step, validation of the model was carried out based on the data of engineering surveys of 2018 (Table 1). The model reproduces water level and flooding areas for all hydrological situations with channel and floodplain roughness coefficient of 0.020 and 0.050, respectively; the differences between the simulated and observed water levels at Blagoveshchensk were less than 10 cm.

An additional validation was carried out for the outstanding floods of 1984 and 2013 using riverbed bathymetry of 2018 surveys. On 15 August 1984, a flood of 1 % exceedance probability took place near Blagoveshchensk; water levels at the station Amur–Blagoveshchensk came up to a maximum height of 868 cm (128.56 m a.s.l.), some streets of the city near the river were inundated. The recent extreme flood of 2013 was characterized by the long period of flooding in the entire Amur basin (Danilov-Danilyan et al., 2014; Bolgov et al., 2015). The maximum water level at the station Amur–Blagoveshchensk during this flood exceeded the height 822 cm above the station zero (128.10 m a.s.l.) on 16 August and the model reproduce this flood with the good accuracy, difference between modelled and observed water levels by the station Amur – Blagoveshchensk is 11 sm, by the station Zeya – Blagoveshchensk is −5 sm. The additional analysis of the inundation zones was performed using satellite images Radarsat and Landsat, the difference between simulated and observed flooded areas was less than 10 %.

The comparison of modelled and observed channel changes was done on the previous step of research for the period 2011–2015 years and has showed, that the model captured location of the main erosion and deposition zones, as well as their amplitude (Krylenko et al., 2018).

5 Results of scenario modelling and their discussion

Results of estimations of water discharges redistributions has showed, that due to the dams construction in the side channels, flow will be concentrated in the main channel of the Amur River under discharges 1500–5500 $m^3 s^{-1}$, flow in the main channel will be increased up to 15 %–20 %. As result, water levels in the main channel of the Amur River near Blagoveshchensk according scenario 2 will be higher on the 15–20 cm in comparison with scenario 1. During peak flow, dams will be overflowed and their influence on the discharges redistribution will be smaller. The water discharges in the main channel of the Amur River above the confluence during peak floods will be increased up to 5 %–6 %, water levels

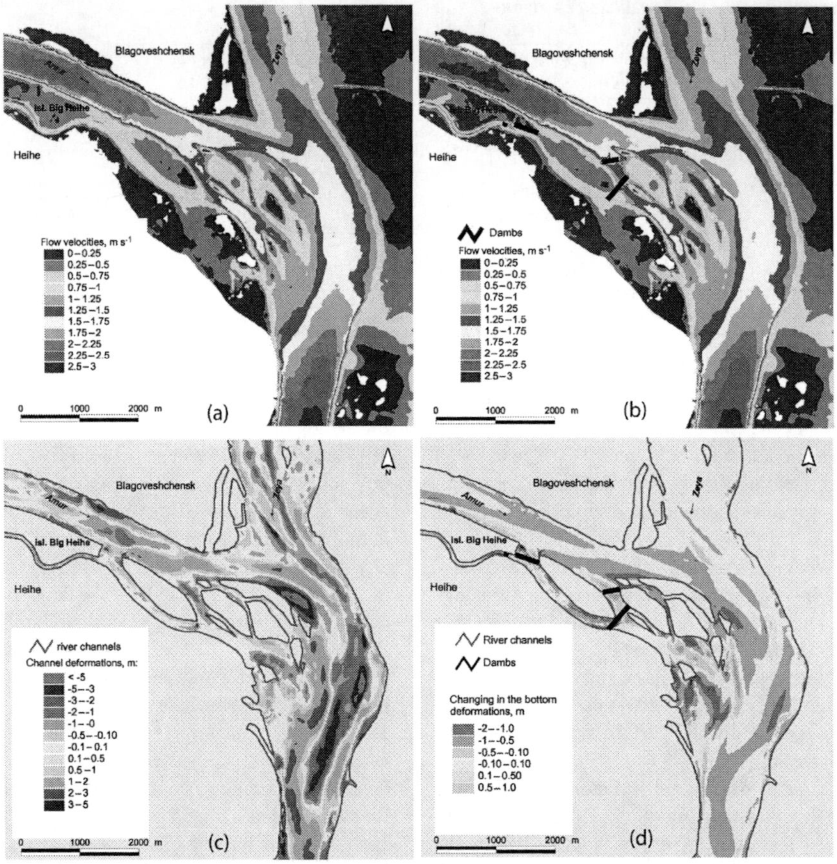

Figure 2. Modelled flow velosities during peak flow in the 2013 year according scenario 1 **(a)** and scenario 2 **(b)**; modelled channel evolution at the confluence of the Amur and Zeya rivers for 5-year forecast period according scenario 1 **(c)**, changes in channel bottom changes due to construction of dams (difference in changes between scenarios 2 and 1) **(d)**.

– up to 10–15 cm. The concentration of the water flow in the main channel will lead to the increasing of the flow velocities near the station Blagoveshchensk on to 0.15 cm s^{-1}, and up to 0.35 cm s^{-1} downstream to the river confluence (Fig. 2).

The channel evolution analysis for the end of 5-year forecast period has showed that location of erosion and depositional areas in the river channels will be same as in the current period for the both scenarios. There are the main erosion zone at the Muraviev Island and in the area at the mainstream line of the Amur River downstream from the confluence and the deposition zone downstream the islands. However, due to the blocking of the side branches by the complex of dams, the erosion in the main channel of the Amur River will increase onto 1–1.5 m compared to the scenario (1) without the dams. Mainstream of the Amur River as well as the Russian – Chinese boundary can shift by 50–70 m toward the left bank within a reach more than 1 km long downstream from the confluence (Fig. 2). The erosion in the main channel can lead to the lowering of the water levels in the end of 5-year forecast period on to about 10 cm. As result the two processes – concentration of the flow in the main channel and channel erosion will be in similar conditions regarding water levels changes.

6 Conclusions

The two-dimensional hydrodynamic model of the important transboundary Russian-Chinese key area at the confluence of the Amur and Zeya rivers was developed based on STREAM_2D software. The model reproduces rather well all hydrological situations including extreme flood of 2013 year and captures the main features of channel processes.

The analysis of the possible influence of the dam construction along channels around the Big Heihe Island on the water flow and channel changes was done using scenario modeling approach.

The comparison of flow characteristics for two scenarios – without dams (1) and with dams (2) has showed that allocation of dams can lead to the increasing of water discharges, water levels and flow velocities in the main channel of the Amur River upstream the confluence. Analysis of channel changes at the end of 5-year forecast period demonstrated 1 m erosion increase at the main channel of the Amur River downstream from the confluence with the Zeya River according scenario (2). However, the water level growth will be less pronounced due to channel erosion, which causes that no significant changes in the flooding zones should be expected.

The further monitoring of situation in the confluence of the Amur and Zeya rivers including projects of constructions on the floodplains and in the channels should be proceed using developed model due to complex influence of natural and anthropogenic factors on the flow characteristics.

Data availability. Data available upon request to the contact author.

Author contributions. VB developed mathematical model, PG prepared digital terrain models, MS analyzed hydrological data, AZ developed experiment design and supervised the project, EF and IK performed the simulations and result analysis.

Competing interests. The authors declare that they have no conflict of interest.

Special issue statement. This article is part of the special issue "Land use and climate change impacts on erosion and sediment transport". It is a result of the ICCE Symposium 2018 – Climate Change Impacts on Sediment Dynamics: Measurement, Modelling and Management, Moscow, Russia, 27–31 August 2018.

Acknowledgements. The authors of the paper express their deep gratitude to the Laboratory of Soil Erosion and Channel Processes of the geographical faculty of Moscow State University and CJSC Lengiprorectrans for providing the field research materials.

Financial support. Analysis of the floods characteristics was supported by the Russian Foundation for Basic Research (grant RFBR no. 17-05-01230), the channel changes evaluation was supported by the Russian Science Foundation (grant RSF no. 18-17-00086).

References

Aleksyuk, A. I. and Belikov, V. V.: Simulation of shallow water flow with areas and bottom discontinuities, Comp. Math. Math. Phys., 57, 318–339, https://doi.org/10.1134/s0965542517020026, 2017.

Belikov, V. V. and Kochetkov, V. V.: Software Complex STREAM_2D to Calculate Streams, Bottom Deformation, and Pollutants Transfer in Open Flows, Software State Registration Certificate no. 612181, Russian Agency for Intellectual Property, 2014.

Belikov, V. V. and Semenov, A. Yu.: Non-Sibsonian interpolation of arbitrary system of points in Euclidean space and adaptive isolines generation, Appl. Numer. Math., 32, 371–387, 2000.

Belikov, V. V., Glotko, A. V., Belousova, I. V., and Zavadsky, A. S.: Application of numerical hydrodynamic modelling for solving border water objects in Siberia, Barnaul: Fundamental'nyye problemy vody i vodnykh resursov [Barnaul: Basic Problems of Water and Water Resources], Water and ecological problems of Siberia and Central Asia, 1, 7–15, 2012 (in Russian).

Belikov, V. V., Krylenko, I. N., Alabyan, A. M., Sazonov, A. A., and Glotko, A. V.: Two-dimensional hydrodynamic flood modelling for populated valley areas of Russian rivers, Proc. IAHS, 370, 69–74, https://doi.org/10.5194/piahs-370-69-2015, 2015.

Bolgov, M. V., Alekseevski, N. I., Gartsman, B. I., Georgievski, V. Yu., Dugina, I. O., Kim, B. I., Makhinov, A. N., and Shalygin, A. L.: The 2013 extreme flood within the Amur basin: analysis of flood formation, assessments and recommendations, Geogr. Nat. Resour., 36, 225–234, 2015.

Danilov-Danilyan, V. I., Gelfan, A. N., Motovilov, Y. G., and Kalugin, A. S.: Disastrous flood of 2013 in the Amur basin: genesis, recurrence assessment, simulation results, Water Resour., 41, 115–125, https://doi.org/10.1134/S0097807814020055, 2014.

Krylenko, I., Belikov, V., Fingert, E., Golovlyov, P., Glotko, A., Zavadskii, A., Samokhin, M., and Borovkov, S.: Analysis of the impact of hydrotechnical construction on the Amur river near Blagoveshchensk and Heihe cities using a two-dimensional hydrodynamic model, Water Resour., 45, 112–121, https://doi.org/10.1134/S0097807818050378, 2018.

Zaitsev, A. A., Belikov, V. V., and Militeev, A. N.: Using computer modeling for a large river, Proc. Int. Symp. "Sediment Transfer through the Fluvial System", Moscow, IAHS Publ. 288, ISBN 1-901502-67-8, 386–394, 2–6 August 2004.

Proc. IAHS, 381, 79–86, 2019
https://doi.org/10.5194/piahs-381-79-2019

Quantifying channel bank erosion of a small mountain river in Russian wet subtropics using erosion pins

Yulia Kuznetsova[1], **Valentin Golosov**[1,2,3], **Anatoly Tsyplenkov**[1,3], **and Nadezhda Ivanova**[1]

[1]Laboratory of Soil Erosion and Fluvial Processes, Faculty of Geography, Moscow State University, GSP-1, Leninskie Gory, 119991, Moscow, Russia
[2]Kazan Federal University, Kremlevskaya st., 18, 420008, Kazan, Russia
[3]Institute of Geography RAS, Staromonetny per., 18, 119017, Moscow, Russia

Correspondence: Yulia Kuznetsova (kuzyulia@gmail.com)

Published: 1 August 2019

Abstract. Channel bank erosion was monitored using erosion pin arrays in a small mountain stream located in the Tsanyk River basin near the Caucasian Black sea coast. The study area is characterized by high precipitation, low mountainous topography, and flysch bedrock. Erosion pins arrays were installed at six key sections each with different bank lithology and morphology. Bank lithology (bedrock or alluvial deposits), morphology (vertical or concave) and orientation towards the local flow direction were the dominant factors controlling the rate of bank retreat and/or collapse. In contrast to alluvial banks, bedrock slopes are predominantly affected by chemical weathering processes in a wet subtropical climate. Results on sediment mobilization and volume removed are presented for different bank types. Variability in channel bank recessional dynamics is discussed in the context of hydrometeorological data (2016–2018) and an extreme flood.

1 Introduction

River channel bank erosion is controlled by a complex combination of factors and processes that play a key role in fluvial dynamics (Hooke, 1979). Channel bank retreat involves a combination of three primary mechanisms, which include subaerial weathering, mass failure and fluvial erosion (Thorne, 1982). Other processes such as creep and seepage erosion can be important under certain situations (Lawler, 1993). The wet subtropical area in Russia consists of low relief topography along a narrow strip of the Caucasian Black Sea coast (Fig. 1). This region is extremely popular among tourists because of its climate, topography and proximity to the sea and the region has experienced considerable growth of infrastructure in recent decades. This process has been especially triggered by the recent Winter Olympic Games (Sochi-2014) and a set of Formula-1 races during the following years. Because the spatial extent of marine terraces is limited in this region, infrastructure development is increasingly located on narrow floodplains and steep slopes of valleys that dissect the low mountain belt of the Caucasus Mountains. Mountainous topography and subtropical climatic conditions in combination with easily weathered flysch-dominated bedrock lead to an increased number of geomorphic hazards. According to local authorities, up to 85 % of the slopes in the Sochi administrative area are susceptible to landslides (Shoygu, 2010).

River bank erosion is a second factor that causes damage to buildings typically located next to the channel. Potential hazards of landslides and fluvial erosion may be exacerbated under conditions of a changing climate (Goudie, 2006; Gariano and Guzetti, 2016). Annual precipitation in this area is over 1500 mm and several heavy rainstorms (100–200 mm per event) can occur each year. During the last decade the number of extreme meteorological events have been observed in the Caucasian Black Sea coast region (Meredith et al., 2015; Alekseevsky et al., 2016) and the frequency of high magnitude rain events has been reported (Magritsky, 2014). Accordingly, in order to mitigate risks due to natural hazards in this region it is necessary to investigate key processes that control channel dynamics and bank erosion rates. In the present study, erosion pin arrays were used to quantify the effects of extreme events river bank recessional rates. This

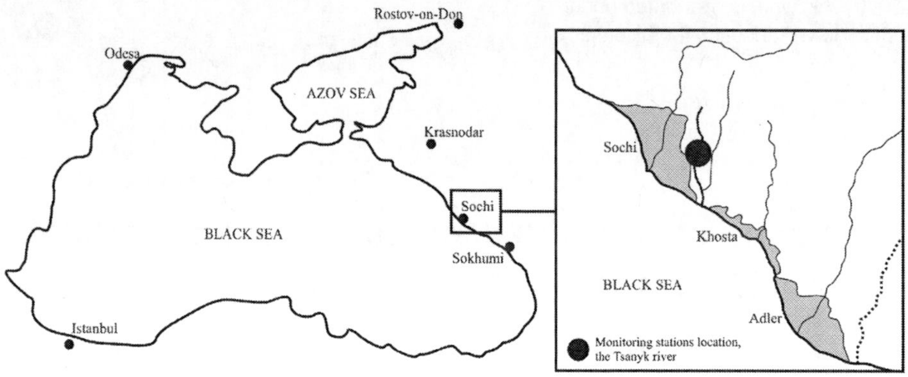

Figure 1. The study site location.

erosion assessment method has been widely applied over the world and advantages and limitations of the method are widely reported (Hooke, 1979; Neller, 1988; Sirvent et al., 1997).

2 Methodology

2.1 Study catchment

The study was conducted in the Tsanyk river valley which is located in Russia near the coast of the Black Sea (Fig. 1). This area is located in the Big Sochi area in the Khosta district and a monitoring program of exogenic processes has been ongoing since the early 2000s (Kuznetsova et al., 2012; Golosov et al., 2012) and the present study is part of river channel process investigations. The Tsanyk river is 12.5 km long and drains an area of 11.7 km^2. The upper section of the river valley consists of a deep V-shaped gorge underlain by solid bedrock unsuitable for land use. The primary geomorphic hazards in this section of the valley are related to slope instability, mainly rock falls due to fieldwork observations. The stepped long profile of the stream is deeply incised while the lower sections of the valley is predominantly urbanized and has undergone significant landscape change.

Along the middle reach of this river which is 2.3 km in length, there is a relatively low level of development which is ideal to evaluate erosion in natural channels to estimate sediment redistribution dynamics within the basin. Detailed geomorphic mapping of the entire river reach was performed and the morphological characteristics of more than over 100 channel and valley bottom cross-sections were measured. Information on relative bank height, width, lithology, vegetation type, and dominant exogenic processes was collected. Survey data were used to select representative channel banks with characteristic lithology, slope and soil characteristics (Fig. 2). Notably, two main types of channel banks were evaluated. They include (1) weathered bedrock consisting of flysch and (2) loose loamy alluvial soil with layers of pebbles and other alluvial debris. Fluvial processes such as particle detachment by flowing water are also accompanied by another complex set of other processes that include mass wasting and weathering (Thorne, 1982). Weathering is the process, which constantly affects all types of river banks, but it plays an especially significant role for bedrock banks by preparing the loose material for further removal by a flow.

2.2 Bank Erosion Estimation

Based on a geomorphic survey of the studied river reach, six sections were selected for monitoring of channel banks dynamics. All sites are located along eroded non-vegetated banks but differ in terms of lithology, micromorphology, and orientation towards the stream (Fig. 3). The main characteristics of the sections are listed in Table 1.

In September 2016 erosion pins (20 cm long steel nails) were deployed at all monitoring sections. All pins were arranged in vertical rows, mostly in checkerboard order. The average distance between neighboring pins ranged from 40 to 60 cm. This density is used to minimize artificial impacts on slope stability, while allowing coverage of all bedrock and soil layers. The exposed parts of the pins were measured in the field with an accuracy of 0.1 cm.

Further measurements were carried out three times: in September 2017, and twice in September 2018 (on 4 September, and right after one extreme flood on 9 September). These measurements allowed changes to channel bank morphology and volumes of the eroded material to be calculated.

In case the pin was not presented at the site, the reason was determined in the field as either erosion to the full length of the pin, or accumulation on top of the pin. In unclear cases we used photographs taken before pins installation and during the repeated measurements to compare local conditions and slope morphology.

For each monitoring section, the map of erosion pins in plan projection was prepared using the GIS software package MapInfo. Based on repeated measurements, a digital sur-

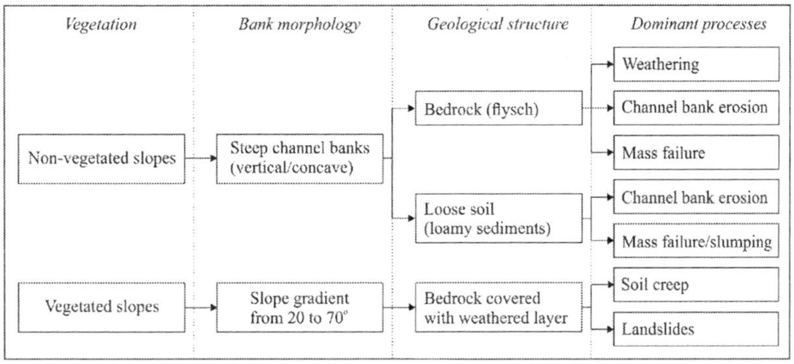

Figure 2. Channel bank types with major characteristics.

Table 1. Monitoring river bank section characteristics.

Monitoring section	Number of pins	Lithology	Section area (m^2)	Bank profile morphology
1	34	Loose soil	6.5	Vertical slope
2	10	Flysch	1.4	Concave slope
3/1	10	Flysch	1.5	Concave slope
3/2	5	Flysch	0.2	Concave slope
4	16	Loose soil	1.5	Vertical slope
5	24	Loose soil	3.4	Vertical slope
6	22	Loose soil	2.3	Vertical slope

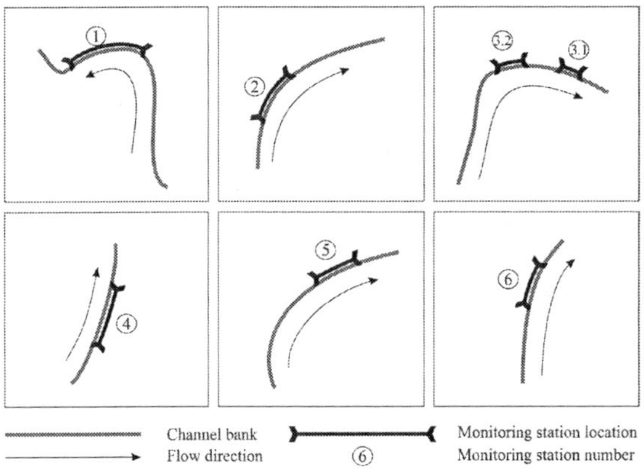

Figure 3. Locations of monitoring sections on the river banks relatively to flow directions.

face model (DSM) and 3-D-model of micro-morphological changes were evaluated for each site and each observation date. The scheme of bank deformation at the monitoring section 6 between 2016 and 2017 is shown in Fig. 4 as an example. Positive values correspond to erosion and negative values correspond to local deposition at the bottom of the eroded bank. In cases where some pins were lost, there were three ways of slope morphology changes calculations. For eroded pins, we used the minimum possible value (its full length of 20 cm). When there was deposition on top of the pin we used zero value, which definitely leads to underestimation of sediment redistribution within the particular bank section, but does not affect total erosion rates calculations and helps to keep the sediment in place for further observations. In cases where no reason was established for the missing pin, it was excluded from further calculations. The input of other processes accompanying fluvial erosion (such as mass failure and weathering) is included in calculation results by default.

We used the field measurement of the channel banks height and length of the eroded bank segments for two bank types based on lithology. Sediment volume eroded from the mass movement-dominated slopes are calculated using the soil creep rate and the thickness of the creep-affected weathered layer. For the other channel bank types, we used the average values obtained at the monitoring stations.

2.3 Hydrometeorological data collection

Meteorological data were collected at the station, located at Razdolnoe village within the river basin, near the western drainage basin divide of the Tsanyk catchment. Streamflow was measured using automatic level gauge (43.57661, 39.78230) connected to an automatic system of flood monitoring in Krasnodar region (http://www.emercit.ru/map/, last access: 15 September 2018). This system has provided open

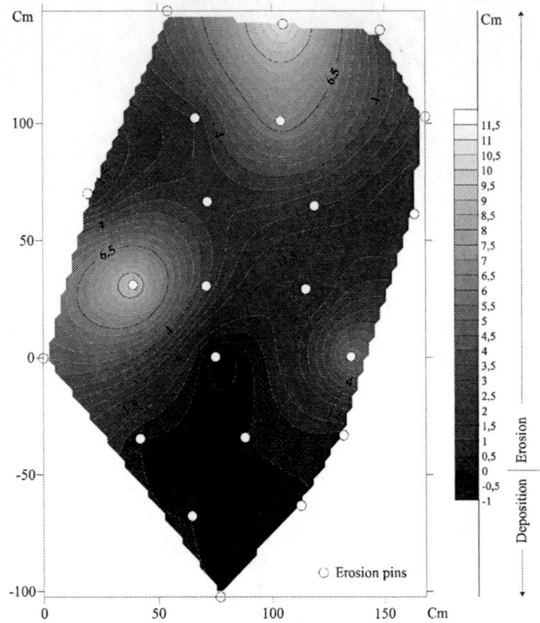

Figure 4. Example map of the channel bank morphology changes. Monitoring section 6, 2016–2017.

source data since 7 April 2014, with 10 min time resolution. Water level assessment accuracy is 1 mm. To avoid artificial peaks and sensor noise the data was corrected manually according to the method of Rodda and Little (2015).

3 Results and discussion

The graphs of significant meteorological events and their hydrological responses are shown at Fig. 5. It is clear that most severe rainfall events happen during wet winter season lasting from October till March. Water level fluctuations correspond well to precipitation events.

The two graphs comparison shows some special characteristics of the hydrological response. The same precipitation layer may cause either an extreme flood or almost no water stage rise. Duration of rainfall and the antecedent dry period play the most important role in the response type. In many cases the longer the period of rains, the higher the water level at its end.

Speaking of the two years of observations, there are significant differences in precipitation distribution by seasons and single events. The year 2018 is obviously wetter. It has 1 month longer wet season, with several notable rainfalls in summer, and the autumn rains started early in September with an extreme hydrological and meteorological event (Fig. 6).

Table 2. Mean volumetric bank retreat rates for different monitoring intervals based on application of erosion pin method.

Monitoring section	Volume of displaced material ($m^3/10\,m^2$)		
	1 Sep 2016– 5 Sep 2017	5 Sep 2017– 4 Sep 2018	4 Sep 2018– 9 Sep 2018
1	0.27	0.76	1.0
2	0.09	0.02	0.28
3/1	0.24	0.50	0.27
3/2	0.08	−0.16	0.18
4	0.30	1.12	0.20
5	0.42	−0.14	–
6	0.41	−0.07	0.14

There was a 43 mm rainfall layer in the morning of 7 September, and the water level rose up to 36 cm. Additional 23 mm of rainfall over the next morning caused the 117 cm water level rise during 40 min only. Such a high and sharp peak is unique for at least the last 5 years – since the start of water level monitoring at this river. Moreover, it was caused by a regular event (23 mm is the 99.9 % frequency rain) that happened approximately 24 h after 90 % frequency rain (43 mm). As mentioned above, there is no direct correlation between precipitation rate and water level, and even 25 % or lower recurrence interval rainfall events did not cause such consequences over the last 5 years. We believe that the antecedent dry period duration and certain sequence in rainfall events are the main reasons for this phenomena.

We were lucky in the field to make measurements of erosion pins on all monitoring sections a few days before and immediately after this unique flood event and to assess the input of the extreme event into the channel banks dynamics.

Estimates of bank erosion are presented in Table 2 and Fig. 7. All volumes of displaced material are given in common volumetric denominator (m^3 of removal or gain over $10\,m^2$ of the channel bank area).

The data show erosion rates both within the single bank section and along the whole river reach, are highly variable depending upon the nature of the bank materials and streamflow. Erosion rates where typically higher in banks composed of loose alluvial material but a number of other factors affecting bank erosion were observed. The bank orientation towards the flow direction plays an important role in the erosion process, which in some cases is even more important than lithology of the eroded banks. The most unstable part of the slope with loose banks is usually located at the meander apex and at bank segments oriented perpendicular to the flow direction.

Bedrock banks are characterized by lower variability of erosion rates than those composed of alluvial sandy-loamy material, apparently more influenced by local factors, such as microscale differences in lithology, morphology, and, mainly, orientation towards the stream flow direction.

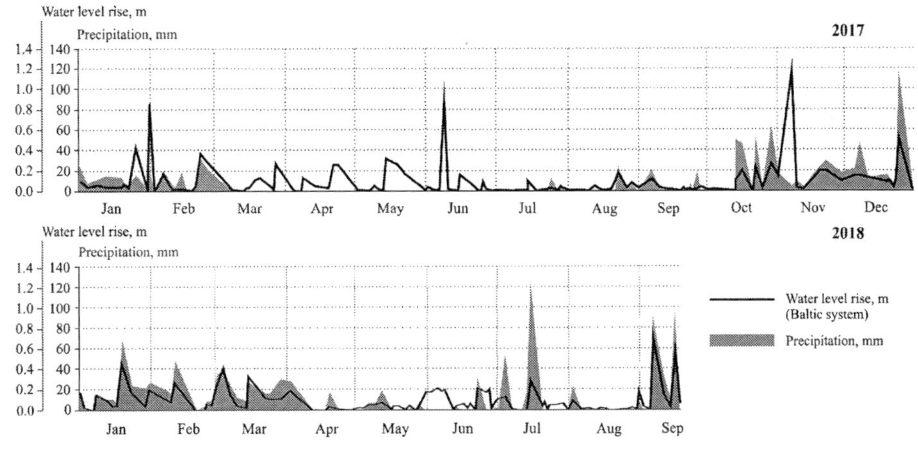

Figure 5. Water level (1) and precipitation (2) changes during the monitoring period.

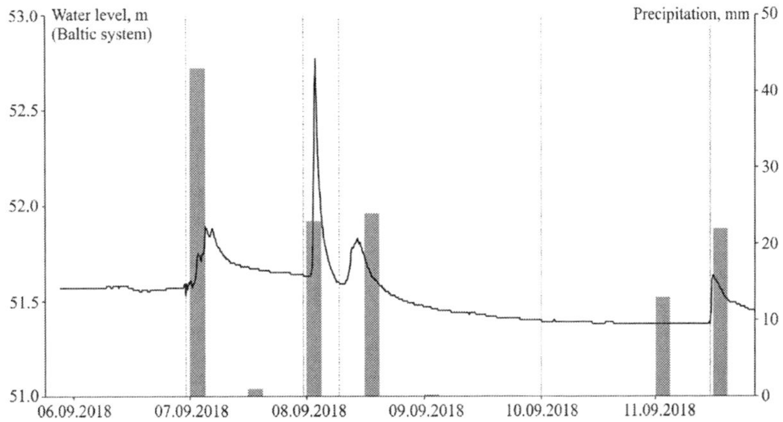

Figure 6. Hydrological events and rainfall in September 2018.

The noticeable exceptions are sections 5 and 6 where the floods of 2018 led to the mass failure of all the channel banks, though the fate of collapsed material depended on the bank orientation towards the channel. The monitored banks at sections 5 and 6 are located in parallel to the stream direction (see Fig. 3), and it takes longer to erode collapsed blocks and flush the produced sediments away. The monitoring results show temporary deposition of the collapsed material along the bank toes at sections 5 and 6.

At station 3/2 there is an accumulation of talus and small rockfall debris which is located directly where the flow hits the bank (see Fig. 3). We speculate that the main reason for such variability is related to bedrock weathering and the mechanisms of the processes following the erosion events. At this site, erosion of the bottom sections of the banks has triggered the weathered banks to collapse. The lag between bank undercutting and mass failure in such cases can take longer than banks consisting of loose sediments, where slumping and sliding usually immediately follow the undercutting and often occur during the same flood. That is why fluvial erosion at the bedrock banks may be delayed comparatively to the other eroded bank sections. In many cases, failures of bedrock banks occur after one event and the collapsed blocks will be eroded during the next floods.

The comparison of situations observed on the sections 3/1 and 3/2 shows that the bank segment located at the meander exit and parallel to the stream collapses faster than the upstream part of the meander where the flow contacts the bank directly. Flysch at the bank surface has relatively high erosional resistance and the intensity of bank deformation will largely depend on differential weathering rates. The flow has to take away and replace the prepared (weathered) debris. The 3/2 section is located under and overhanging upper part of the high north facing slope, where weathering rates can

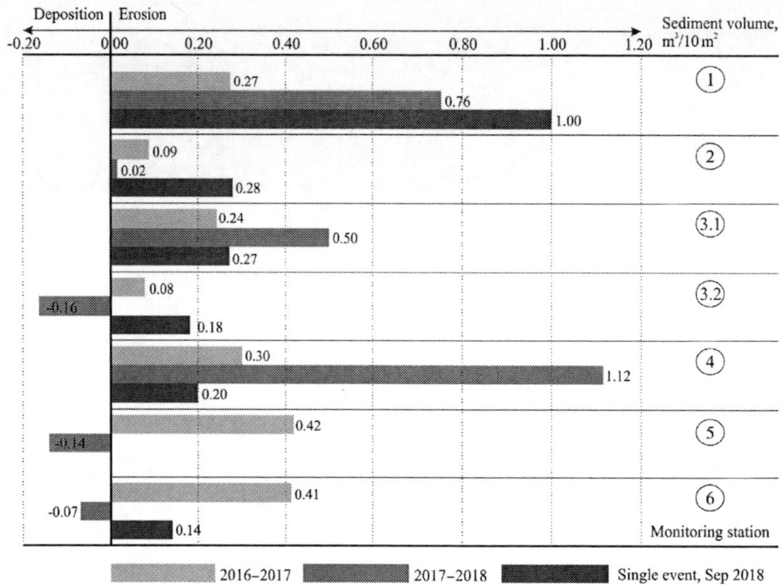

Figure 7. Volumes of mobilized material over the observation periods.

Table 3. Total volumes of displaced material along the studied river reach.

Channel bank type	Total length of all segments, m	Average channel bank relative height, m	Volume of displaced material, m^3		
			1 Sep 2016–5 Sep 2017	5 Sep 2017–4 Sep 2018	4 Sep 2018–9 Sep 2018
Eroded bedrock bank	550	3.5	26.2	22.8	47.3
Eroded loose soil bank	2050	1.8	130.4	153.7	164.7
Vegetated soil creep bank	800	–	1	1	1
Accumulative bank	1100	–	–	–	–

be expected to be lower. In contrast, section 3/1 is more susceptible to the temperature fluctuations, so the rocks collapse faster, and the total volume of mobilized debris is greater.

All monitored banks have significant differences in their different parts. The most exposed erosion pins are usually observed at the middle part of the bank, while the lower parts and the toes are often covered with the collapsed sediments, waiting to be flushed away during the following flood event. The only exception observed is the case of the extreme single flood when almost all debris was immediately removed by the stream.

Comparison of results of two years of observation (not accounting for the extreme event) shows that differences of the channel bank dynamics depend on the amount and type of precipitation. Volumes of the displaced material in 2018 are 1.5–2.5 times greater than during the drier previous year of 2017. A number of hydrological events played a significant role in this process. The contribution of the unique hydrological September 2018 event had a dramatic effect on channel bank erosion dynamics. The largest erosion rates occurred in section 1 which consists of loose alluvial sediments oriented towards the flow direction. The volumes of material removed during this particular event was 25 % higher than during the prior 8 months of the wet year, and almost 4 times higher than the total sediment loss over the drier year of 2017 with no significant hydrological events observed. At all other sections, volumes of material displaced over a single extreme event are at least comparable with the annual bank retreat during the previous period of monitoring.

To calculate the sediment volumes supplied by slow mass movement on vegetated banks, we used the process rate value of 2 mm yr^{-1} which is based on long-term monitoring assessment of the creep rates using the Young method (Young, 1960; Azhigirov and Golosov, 1990) during previous studies at the same area (Kuznetsova et al., 2012). Soil depth was determined to be approximately 0.7 m was used for the sediment delivery calculations.

The total volume of eroded bank material was calculated for the studied river reach for the entire period of monitoring and for a single extreme hydrological event (Table 3).

Total volumes of displaced and mobilized material for the whole 2.3 km long river reach from all types of the banks range from 150 m^3 in the relatively low-water year of 2017 to more than 300 m^3 in 2018 when extreme hydrological events occurred during the year of observation. The contribution of one of such events to the total redistribution of coarse and fine sediments along the river may be equal to the annual bank retreat. These preliminary estimates, provide useful information on the contribution of within-channel sources in the total sediment load of the Tsanyk River.

4 Conclusions

Monitoring of channel bank retreat using the erosion pin method shows the high general intensity of erosion and associated processes at the wet subtropics. The sediment erosion rates were highly variable depending on local morphology, hydro-climatological and weathering conditions within banks of differing lithology. The main drivers of bank erosion are lithology and its microscale variations, weathering, slope morphology, and bank orientation towards the flow direction. The contribution of these factors may vary substantially, and they all may play the main role in channel bank dynamics depending on local conditions. In general, banks composed of loose alluvial sediments are eroded faster than bedrocks, though in some cases the rates of banks collapses and volumes of mobilized coarse and fine sediments may be comparable. For the latter cases weathering especially active under conditions of wet subtropical climate becomes the leading process.

Volumes of the mobilized material (and the resulting stream sediment load) are highly dependent on the local meteorological and hydrological situation. We suggest that antecedent conditions of the relatively prolonged dry period prior to the most significant hydrological events are the most favorable conditions for extreme channel banks retreat or collapse. Results of the study have practical applications for engineering projects within the valley, evaluation of bank erosion-imposed hazards for different structures along the Tsanyk river channel, as well as for similar small valleys typical for the entire Caucasian Black Sea coast.

Data availability. Pins measurements dataset is available at public data repository. Please, use the link https://zenodo.org/record/3332732#.XSc8xfZuI0Q (last access: 11 July 2019). For more informantion contact Yulia Kuznetsova (kuzyulia@gmail.com).

Author contributions. YK, VG and NI conceived and designed the research, YK and AT analyzed the data, YK, VG and AT edited the manuscript.

Competing interests. The authors declare that they have no conflict of interest.

Special issue statement. This article is part of the special issue "Land use and climate change impacts on erosion and sediment transport". It is a result of the ICCE Symposium 2018 – Climate Change Impacts on Sediment Dynamics: Measurement, Modelling and Management, Moscow, Russia, 27–31 August 2018.

Acknowledgements. This article was prepared with financial support from the Russian Fund of Basic Researches, project no.16-05-00815. The authors wish to thank Dmitry Botavin for the valuable help during fieldwork.

Financial support. This research has been supported by the Russian Fund of Basic Researches (grant no. 16-05-00815).

References

Alekseevsky, N. N., Magritsky, D. V., Koltermann, P. K., Toropov, P. A., Shkolny, D. I., and Belyakova, P. A.: Floods at the Black Sea coast, Krasnodar region, J. Water Resour., 43, 3–17, https://doi.org/10.7868/S032105961601003X, 2016.

Azhigirov, A. A. and Golosov, V. N.: Estimation of slow mass movement for civil engineering investigations, J. Geomorphologiya, 1, 33–40, 1990.

Gariano, S. L. and Guzetti, F.: Landslides in a changing climate, Earth-Sci. Rev., 162, 227–252, https://doi.org/10.1016/j.earscirev.2016.08.011, 2016.

Golosov, V. N., Delia Seta, M., Azhigirov, A. A., Kuznetsova, Y. S., Del Monte, M., Fredi, P., Lupia Palmieri, E., and Grigorieva, T. M.: Anthropogenic activity influence on exogenic processes intensity in low mountains of the subtropical zone, J. Geomorphologiya, 2, 7–17, https://doi.org/10.15356/0435-4281-2012-2-7-17, 2012.

Goudie, A. S.: Global warming and fluvial geomorphology, Geomorphology, 79, 384–394, https://doi.org/10.1016/j.geomorph.2006.06.023, 2006.

Hooke, J. M.: An analysis of the processes of river bank erosion, J. Hydrol., 42, 39–62, https://doi.org/10.1016/0022-1694(79)90005-2, 1979.

Magritsky, D. V.: Spatial and temporal characteristics of the floods at the Black Sea coast of Russian Federation, J. Vestnik of Moscow University, 5, Geography, 39–47, 2014.

Kuznetsova, Y. S., Grigoreva, T. M., Golosov, V. N., Markelov, M. V., Derkach, A. A., and Bolysov, S. I.: Monitoring of exogenic geomorphic processes in urban areas of the Sochi Region, the Caucasian Black Sea coast, in: Antropopresja w wybranych sirefach morfoklimatycznych – zapis zmian w rzezbie I osadkach, Sosnowec, Polonia, 226–230, 2012.

Lawler, D. M.: The measurement of river bank erosion and lateral channel change: A review, Earth Surf. Proc. Land., 18, 777–821, https://doi.org/10.1002/esp.3290180905, 1993.

Meredith, E. P., Semenov, V. A., Maraun, D., Park, W., and Chernokulsky, A. V.: Crucial role of Black Sea warming in amplifying the 2012 Krymsk precipitation extreme, J. Nat. Geosci., 8, 615–619, https://doi.org/10.1038/NGEO2483, 2015.

Neller, R. J.: A comparison of channel erosion in small urban and rural catchments, Armidale, New South Wales, J. Earth Proc. Land., 13, 1–7, https://doi.org/10.1002/esp.3290130102, 1988.

Rodda, H. J. E. and Little, M. A.: Understanding Mathematical and Statistical Techniques in Hydrology, John Wiley & Sons, Ltd, Chichester, UK, https://doi.org/10.1002/9781119077985, 2015.

Shoygu, S. K. (Ed.): Atlas of natural and anthropogenic hazards and risks of emergency situations of the Russian Federation. Southern federal district, DIK, Moscow, Russia, 2010.

Sirvent, J., Desir, G., Gutierrez, M., Sancjo C., and Benito, G.: Erosion rates in badland areas recorded by collectors, erosion pins and profilometer techniques (Ebro Basin, NE-Spain), J. Geomorphology, 18, 61–75, https://doi.org/10.1016/S0169-555X(96)00023-2, 1997.

Thorne, C. R.: Processes and mechanisms of river bank erosion, in: Gravel-Bed Rivers, edited by: Hey, R. D., Bathurst, J. C., and Thorne, C. R., Wiley, Chichester, 227–259, 1982.

Young, A.: Soil movement by denudational processes on slopes, J. Nature, 188, 120–122, 1960.

Proc. IAHS, 381, 87–93, 2019
https://doi.org/10.5194/piahs-381-87-2019

Contemporary suspended sediment yield of Caucasus mountains

Anatoly Tsyplenkov[1,2], Matthias Vanmaercke[3], and Valentin Golosov[1,2]

[1]Faculty of Geography, Lomonosov Moscow State University, 119991 Moscow, Russian Federation
[2]Institute of Geography, Russian Academy of Sciences, 119017 Moscow, Russian Federation
[3]Département de Géographie, Université de Liège, Clos Mercator 3, 4000 Liège, Belgium

Correspondence: Anatoly Tsyplenkov (atsyplenkov@gmail.com)

Published: 1 August 2019

Abstract. Processes linked to climate change and intensified anthropogenic pressure influence the environment, the hydrology and by extent the denudation processes in the Caucasus mountain belt. Quantitative assessments of sediment fluxes and their temporal evolution in this mountain region are required for various environmental and engineering purposes, including the planning and maintenance of water reservoirs and other structures. In this paper, we present a first analysis of the hitherto largest suspended sediment yield (*SSY*) database for the Caucasus region, comprising data from 198 catchments (>4000 catchment-years of observations). We present an overview of the existing contemporary *SSY* data from gauging stations observations. Based on these data and different models, we propose preliminary maps of the spatial patterns of *SSY* and denudation rates in the Caucasus region.

1 Introduction

Denudation rates in the mountains are controlled by both endogenic and exogenic factors. Endogenic factors include the degree of tectonic uplift and seismic activity and are typically considered to be the dominant driving force of denudation rates in tectonically active mountain ranges (e.g. Malamud et al., 2004; Montgomery and Brandon, 2002). Nonetheless, also exogenic factors play a crucial role. These include climatic conditions and vegetation cover (e.g. Syvitski and Milliman, 2007). They are especially key factors to understand temporal trends in sediment fluxes and denudation rates at contemporary time scales.

The Caucasus mountains are a highly tectonically and seismically active region, resulting most probably in high long-term denudation rates. At the same time, patterns and trends of climate and vegetation cover likely also exert significant control on suspended sediment yields (*SSY*, [$\mathrm{t\,km^{-2}\,yr^{-1}}$]) at annual to decadal timescales. Disentangling the role of tectonic and geomorphic controls *SSY* from the role of climate and land cover is a central challenge (e.g. Syvitski and Milliman, 2007; Vanmaercke et al., 2015). This is not only of fundamental scientific importance but also of great relevance from a catchment and water resources management perspective (e.g. for the construction and maintenance of water reservoirs). This challenge is especially pertinent in the light of ongoing climate change and increasing anthropogenic pressure (including land use/land cover changes).

A key obstacle in addressing this research challenge is the availability of sufficient and reliable *SSY* observations in order to statistically discern the relative importance of its different driving factors (e.g. Vanmaercke et al., 2011, 2015). Here we aim to help addressing this gap by quantitatively assessing the spatial patterns of contemporary *SSY* in the Caucasus region, based on a large dataset of gauging stations observations from mid the 1920 to 2015.

2 Materials and methods

A large number of studies focussed on compiling and understanding patterns of *SSY* at regional to global scales. Several of these also consider (parts of) the Caucasus region (Dedkov and Moszherin, 1984, 1992; Jaoshvili, 2002; Milliman and Syvitski, 1992; Vanmaercke et al., 2011). We developed a comprehensive database of contemporary *SSY* observations for the Caucasus mountains by integrating these existing datasets and complementing them with hydrological observations at gauging stations in the Russian Federation (State Water Cadastre; https://gmvo.skniivh.ru/, last access: 19 June 2019) as well as *SSY* observations for specific catchments reported by individual studies (Abduev, 2015; Eyubova, 2015; Khmeleva et al., 2000; Magritskii, 2011).

Given that these observations were derived from different sources, also the reliability of the resulting *SSY* varies. This reliability depends on the measuring procedure and especially the sampling frequency (e.g. Moatar et al., 2006). However, also the length of the *SSY* observation periods is an essential factor. Vanmaercke et al. (2012) showed that uncertainties on average *SSY* values based on a short measuring period (1–5 years) due to inter-annual variations in sediment yield are typically at least as large as the uncertainty relating to the measuring period. Moreover, such average *SSY* values are relatively more likely to underestimate the actual average *SSY* (Vanmaercke et al., 2012). Given these concerns, we only considered catchment *SSY* values from direct gauging station (*GS*) observations with a measuring period (*MP*) of at least 10 years.

In total, our Caucasus database comprised *SSY* observations for 198 catchments. An overview of the data is given in Table 1, while the temporal coverage of the database is shown in Fig. 1. The sum of all *MP*s yields a total of 4011 catchment-years of observations. However, for several entries from Dedkov and Moszherin (1984) start and end date was unknown. We assumed a value of 10 years for such entries. With respect to catchment sizes, 17 *GS*s have a catchment area $< 100\,\mathrm{km}^2$, 101 *GS*s have a catchment area between 100 and $1000\,\mathrm{km}^2$, 64 between 1000 and $10\,000\,\mathrm{km}^2$, and 16 have a catchment area $> 10\,000\,\mathrm{km}^2$ (see Fig. 2).

Overall, SSY values ranged between 3.6 and $4100\,\mathrm{t\,km^{-2}\,yr^{-1}}$, with an average value of $442\,\mathrm{t\,km^{-2}\,yr^{-1}}$ and a median of $197\,\mathrm{t\,km^{-2}\,yr^{-1}}$. Two catchments in our database had an exceptionally low *SSY* value (respectively 3.6 and $3.9\,\mathrm{t\,km^{-2}\,yr^{-1}}$). These two catchments were located in the submontane zone of Azerbaijan and Armenia. Given that these *SSY* are clearly lower than what can be expected for this kind of environments (e.g. Vanmaercke et al., 2011) and the reliability of these observations could be questioned, these two observations were not considered for further analyses. According to a Shapiro-Wilks test, the SSY observations are log-normally distributed. All further analyses were therefore conducted on the $\log_{10}$ of the *SSY*.

In order to visualize and better understand the patterns of *SSY* in the Caucasus, we used several methods to spatially interpolate $\log_{10}$ (*SSY*). These included Ordinary Kriging (*OK*), Co-Kriging with an elevation as a co-variable (*CK-h*) and Co-Kriging with a Mean Local Relief as a co-variable (*CK-mlr*), *k*-nearest neighbors (*KNN*) and *k*-nearest neighbors with a co-variable (*KKN-h, KKN-mlr*). *OK, CK, CK-h* and *CK-mlr* were performed with the R package *automap* (Hiemstra et al., 2009), *KNN, KKN-h,* and *KKN-mlr* with the

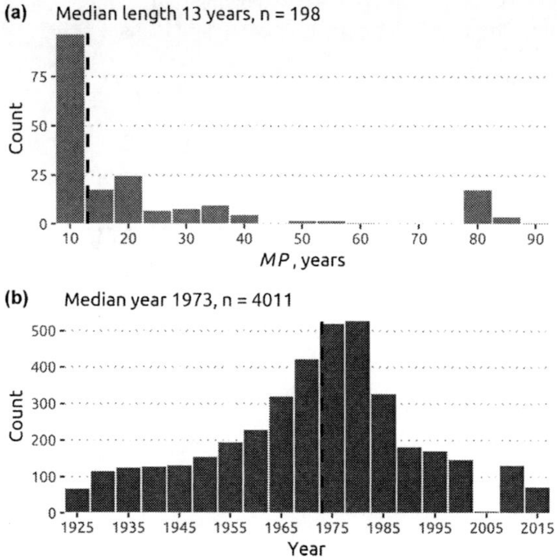

Figure 1. Temporal coverage of the Caucasus *SSY* database. A: distribution of the measuring periods for which gauging station *SSY* observations are available. B: amount of catchment-year *SSY* data available per 5-year period.

R package *kknn* (Schliep and Hechenbichler, 2016). Elevation (*H*, [m]) and Mean Local Relief (*MLR*, [m]) were chosen as an auxiliary variables due to their expected correlation with *SSY* (Ahnert, 1970; Milliman and Syvitski, 1992; Montgomery and Brandon, 2002). Elevation data was accessed with the *elevatr* package (Hollister and Tarak Shah, 2017) in *R*. *MLR* was calculated as a difference between maximum and minimum *H* within a 10 km buffer as it was suggested by Montgomery and Brandon (2002). We randomly split our database into two datasets: one for training and one for validation (Table 2).

The performance of these interpolations was assessed based on the root mean square error (*RMSE*), the mean error (*ME*), the Nash–Sutcliffe efficiency (*NSE*; (Nash and Sutcliffe, 1970), the coefficient of determination (R^2) and Spearman's coefficient of correlation (r).

To assess the total sediment load of the Caucasus, the sediment export of a catchment (*SSL*, $\mathrm{t\,km^{-2}\,yr^{-1}}$) was calculated as:

$$SSL = SSY \cdot A, \tag{1}$$

with: A – the catchment area [km^2]. If there were several *GS* at one river we considered the most downstream station (with the largest catchment area) for calculating *SSL*. For the ungauged basins we made estimate based on our interpolation. We calculate an average *SSY* for every ungauged basin draining to the exit of our study area and multiply this with the size of the basin in order to have an estimate of the *SSL*. To

Table 1. SSY database overview (with outliers).

Country	Number of catchments (catchment – years)	Data sources
Abhazia (part of Georgia)	6 (54)	Khmeleva et. al. (2000), Dedkov and Mozherin (1984)
Armenia	25 (NA)	Dedkov and Mozherin (1984)
Azerbaijan	54 (466)	State water cadastre; Eyubova (2015), Abduev (2015)
Georgia	22 (NA)	Dedkov and Mozherin (1984)
Russia	91 (3491)	Magritskii (2011); State water cadastre;
TOTAL	198 (4011)	https://gmvo.skniivh.ru/ (last access: 19 June 2019)

NA: not available.

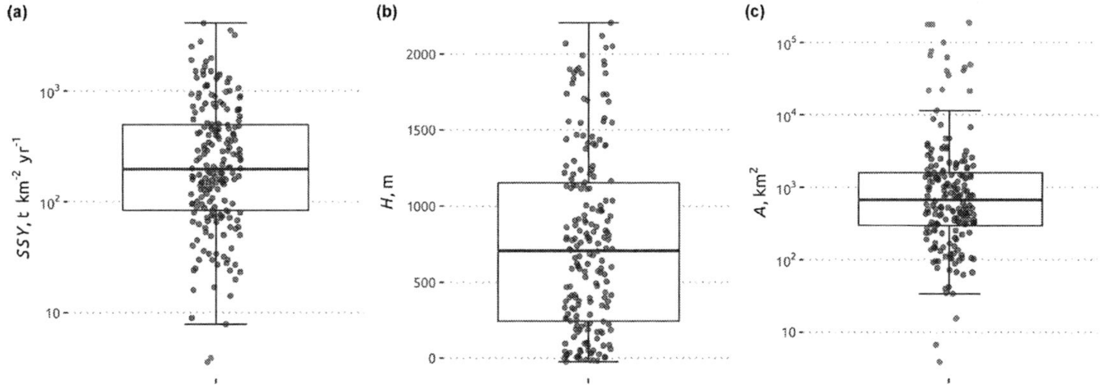

Figure 2. Distribution of *SSY* observations in our database **(a)** and boxplots of elevations of the gauging stations **(b)** and their catchment areas **(c)**.

define basin borders for the ungauged watersheds we used HydroBasins level 5 (Lehner and Grill, 2013).

3 Results

The results of the six interpolation methods (*OK*, *CK-h*, *CK-mlr*, *KNN*, *KNN-h*, *KNN-mlr*) are shown in Table 3. Overall, these results show that observed *SSY* values are strongly spatially correlated. As such, about 70 % of the observed spatial variation in *SSY* can be estimated based on a spatial extrapolation of observations from other *GS*. This is shown by both the *KNN* and *OK* approach. Adding the considered covariable (elevation or Mean Local Relief) slightly improves the prediction quality, but also significantly changes the simulated spatial pattern of *SSY*. The *CK* approach shows the best training results in terms of *NSE*, R^2 and r values but clearly performs worse in terms of validation. The *KNN-h* approach shows the smallest difference between training and validation metrics: both *NSE* and R^2 are above average. Moreover, *KNN-h* has the best results for validation. We, therefore, considered *KNN-h* as the optimal approach to assess the spatial variation of *SSY* in the Caucasus region. Figure 3 illustrates the simulated spatial distribution of *SSY* and elevation.

Overall, mean *SSY* values are higher in the northern part of the Caucasus ($504\,\mathrm{t\,km^{-2}\,yr^{-1}}$) are higher than for the southern part ($396\,\mathrm{t\,km^{-2}\,yr^{-1}}$). The difference is significant and not related to differences in altitude (Fig. 3). Furthermore, *SSY* values are very low ($<50\,\mathrm{t\,km^{-2}\,yr^{-1}}$) in most of Armenia and Azerbaijan, especially in the highland steppic zone. On the other hand, the highest values of *SSY* ($>1000\,\mathrm{t\,km^{-2}\,yr^{-1}}$) are observed in the Dagestan steppes (eastern part of the Caucasus). Furthermore, *SSY* values are significantly higher in the high mountain belt of the central part of Caucasus. This is likely due to the presence of glaciers and the sparse vegetated cover in these headwater catchments.

We estimated that the total measured suspended sediment load (*SSL*) from rivers draining the Caucasus Mountains is about $98\,\mathrm{Mt\,yr^{-1}}$. This value is based on direct measurements from 196 *GS*, which covers 61 % of the total area of

Table 2. Ranges and characteristics of the observed suspended sediment yields and the generated training and validation datasets (without outliers).

	Number of catchments	SSY_{min}, $t\,km^{-2}\,yr^{-1}$	SSY_{max}, $t\,km^{-2}\,yr^{-1}$	SSY_{mean}, $t\,km^{-2}\,yr^{-1}$	SSY_{med}, $t\,km^{-2}\,yr^{-1}$	SD
Train	156	7.9	4100	470	204	660
Validation	40	17	1650	354	164	400
TOTAL dataset	196	7.9	4100	446	200	620

Table 3. Performance measures of the different tested interpolation methods (see text for further details).

Model	Type	RMSE	ME	NSE	R^2	r
KNN	Train	0.31	0.0055	0.7	0.69	0.84
	Validation	0.27	−0.033	0.7	0.7	0.87
OK	Train	0.24	0.00056	0.81	0.81	0.91
	Validation	0.27	−0.046	0.69	0.68	0.86
KNN-h	Train	0.3	−0.0076	0.71	0.7	0.85
	Validation	0.26	−0.013	0.72	0.71	0.84
CK-h	Train	0.24	−0.0002	0.82	0.82	0.91
	Validation	0.27	−0.046	0.7	0.69	0.84
KNN-mlr	Train	0.3	0.00036	0.72	0.71	0.86
	Validation	0.29	−0.068	0.65	0.66	0.83
CK-mlr	Train	0.23	−0.0008	0.83	0.83	0.92
	Validation	0.28	−0.05	0.68	0.67	0.86

the study area. *SSL* of the ungauged part of the Caucasus is $50\,Mt\,yr^{-1}$ (based on the *KNN-h* interpolation).

4 Discussion

The *SSY* database presented in this article is the hitherto largest in its kind. Overall, *SSY* estimates for Caucasus rivers remain relatively scarcely reported in scientific literature. Overall, the average *SSY* of our database ($446\,t\,km^{-2}\,yr^{-1}$) corresponds very well to the average *SSY* reported for catchments in Alpine zones in Europe ($451\,t\,km^{-2}\,yr^{-1}$; Vanmaercke et al., 2011).

We also compared our findings with earlier estimates of denudation rates in the Caucasus. One of the first assessment was made by Gabrielyan (1971), who reported a mean annual denudation rate of Caucasus of $0.2\,mm\,yr^{-1}$. Mozzherin and Sharifullin (2015) estimated that mean annual denudation rates in the Caucasus range between 0.005 and $2.32\,mm\,yr^{-1}$. They calculated denudation rates (h_c, [$mm\,yr^{-1}$]) using the following formula:

$$h_c = \frac{SSY}{2.65} \cdot 10^{-3}, \tag{2}$$

To compare the results of our database with these previous findings, we converted our *SSY* values to h_c using the same equation. As such, the mean denudation rate for the Caucasus region based on our database is $0.17\,mm\,yr^{-1}$ with a maximum h_c of $1.55\,mm\,yr^{-1}$. As indicated by Fig. 4, the spatial patterns of our database correspond relatively well with those estimated by Mozzherin and Sharifullin (2015). The main difference between their and our estimates occurs in the center of the Caucasus (Georgia). The density of *GS* is scarce in this region and, as such uncertainties are especially important in this part of the study area. Areas having a relatively low denudation rate (e.g. $h_c < 0.1\,mm\,yr^{-1}$) agree well for both approaches. However, areas with a high h_c ($> 0.25\,mm\,yr^{-1}$) occupy 31 % of the study region according to our assessment and only 18 % according to (Mozzherin and Sharifullin, 2015). These differences are likely attributable to different measuring periods, differences in interpolation methods, as well as the larger amount of observations used in our study. As such, we believe that our assessment provides a refinement of the earlier estimated spatial patterns of *SSY* for the Caucasus region. Nonetheless, further improvement remains possible. As a next step, we aim to include additional relevant variables (e.g. relating to land cover, climatic conditions, lithology and soil characteristics, topography, tectonics and seismicity) in our assessment of *SSY*. This may not only improve our understanding of spatial and temporal patterns of *SSY* but will also help in disentangling the relative impor-

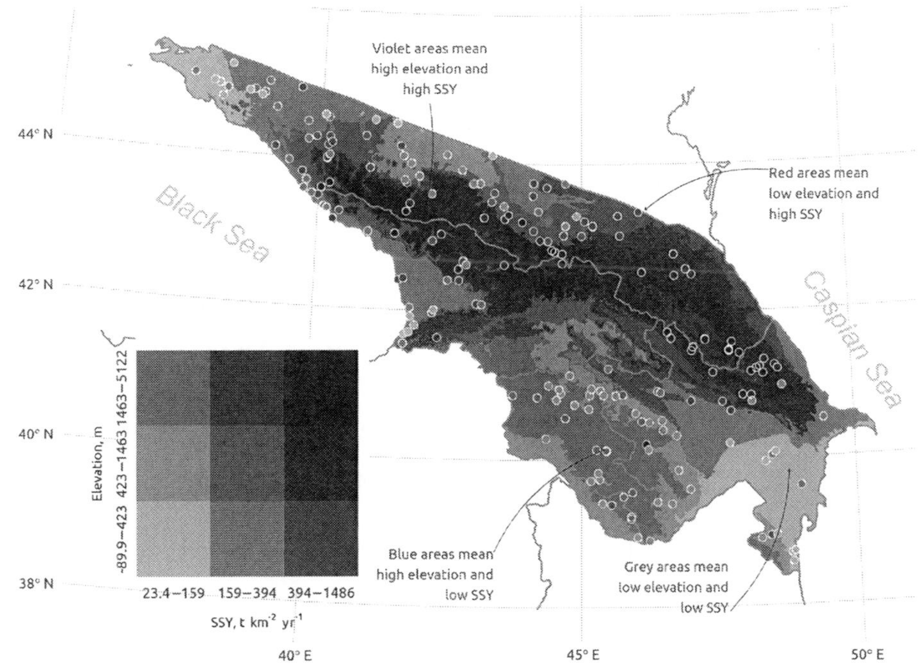

Figure 3. Elevation (H, m a.s.l.) and average annual Suspended Sediment Yields (SSY, t km^{-2} yr^{-1}) as estimated based on our *SSY* database.

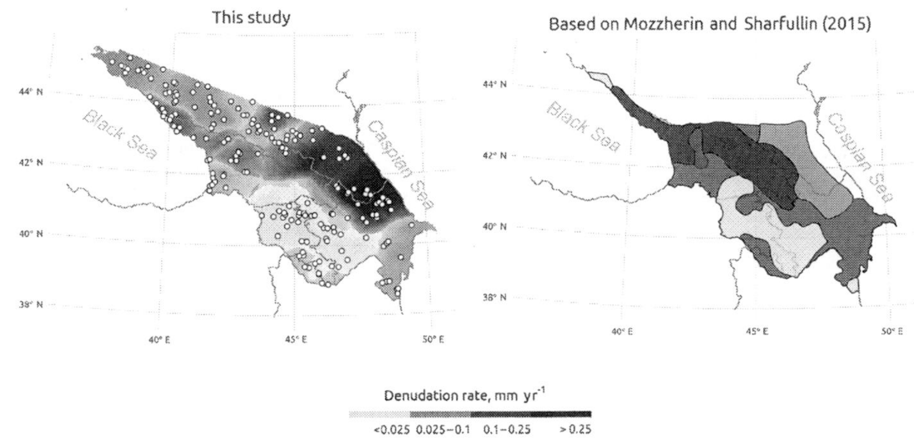

Figure 4. Spatial patterns of estimated average denudation rates for the Caucasus (mm yr^{-1}) and comparison with the estimates of Mozzherin and Sharifullin (2015).

tance of endogenic and exogenic drivers of catchment-wide denudation.

5 Conclusions

In this paper, we presented the hitherto largest *SSY* database for the Caucasus region as well as some first explorative analyses of the spatial patterns of *SSY*. We found that Caucasus *SSY* values are similar in range and average to those of catchments in European alpine climatic zones. Observed *SSY* values are strongly correlated in space, allowing a reasonable assessment of spatial patterns of *SSY* based on a geographical interpolation of *GS* observations. Adding information on elevation further improved the accuracy of our simulations. This methodological approach in combination with the hitherto largest *SSY* dataset allowed to make an updated assessment of spatial patterns of *SSY* in the Caucasus (Fig. 3).

Data availability. Reproducible R code is available at the GitHub repository (https://github.com/atsyplenkov/caucasus-sediment-yield) (Tsyplenkov et al., 2019). Contact Anatoly Tsyplenkov (atsyplenkov@gmail.com) for more information.

Author contributions. VG, AT conceived and designed the research; AT performed the experiments and analyzed the data; AT, VG and MV wrote and edited the manuscript.

Competing interests. The authors declare that they have no conflict of interest.

Special issue statement. This article is part of the special issue "Land use and climate change impacts on erosion and sediment transport". It is a result of the ICCE Symposium 2018 – Climate Change Impacts on Sediment Dynamics: Measurement, Modelling and Management, Moscow, Russia, 27–31 August 2018.

Acknowledgements. This study contributes to the State Task no. 0148-2019-0005, Institute of Geography RAS. The authors wish to thank Oleg P. Yermolaev for sharing an electronic version of the Dedkov and Moszherin, 1984 database.

Financial support. The work (analysis and results) was supported by the Russian Science Foundation (project number 19-17-00181).

References

Abduev, M. A.: Amending the runoff suspended sediment and turbidity of the mountain rivers of Azerbaijan under the influence of economic activity, Geogr. Bull., 32, 29–37, 2015.

Ahnert, F.: Functional relationships between denudation, relief, and uplift in large, mid-latitude drainage basins, Am. J. Sci., 268, 243–263, https://doi.org/10.2475/ajs.268.3.243, 1970.

Dedkov, A. P. and Moszherin, V. I.: Erosion and Sediment Runoff on the Earth, Kazan: Izd-vo Kazansk. un-ta., 264 pp., 1984.

Dedkov, A. P. and Moszherin, V. I.: Erosion and sediment yield in mountain regions of the world, Eros. Debris Flows Environ. Mt. Reg., 209, 29–36, available at: http://iahs.info/redbooks/a209/iahs_209_0029.pdf (last access: 19 June 2019), 1992.

Eyubova, F. A.: The investigation of long-term changes of hydrological regime and of runoff of suspended loads of the rivers of north-eastern slope of greater caucasus (within of Azerbaijan republic), Geogr. Bull., 2, 32–38, 2015.

Gabrielyan, G. K.: The intensity of denudation in the caucasus, Geomorphol. RAS, 1, 22–27, 1971.

Hiemstra, P. H., Pebesma, E. J., Twenhöfel, C. J. W., and Heuvelink, G. B. M.: Real-time automatic interpolation of ambient gamma dose rates from the Dutch radioactivity monitoring network, Comput. Geosci., 35, 1711–1721, https://doi.org/10.1016/j.cageo.2008.10.011, 2009.

Hollister, J. and Tarak Shah: elevatr: Access Elevation Data from Various APIs, available at: http://github.com/usepa/elevatr (last access: 19 June 2019), 2017.

Jaoshvili, S.: The rivers of the Black Sea, EEA Tech. Rep., 71, 58 pp., 2002.

Khmeleva, N. V., Vinogradova, N. N., Samoilova, A. A., and Shevchenko, B. F.: Mountain river basin and exogenous processes within it, edited by: Chalov, R. S., LMSU Faculty of Geography, Moscow, 186 pp., 2000.

Lehner, B. and Grill, G.: Global river hydrography and network routing: baseline data and new approaches to study the world's large river systems, Hydrol. Process., 27, 2171–2186, https://doi.org/10.1002/hyp.9740, 2013.

Magritskii, D. V.: Variations in sediment runoff in the lower reaches and the delta of the Kuban River in the XX–XXI centuries, Water Resour., 38, 709–719, https://doi.org/10.1134/S0097807811060108, 2011.

Malamud, B. D., Turcotte, D. L., Guzzetti, F., and Reichenbach, P.: Landslides, earthquakes, and erosion, Earth Planet. Sc. Lett., 229, 45–59, https://doi.org/10.1016/j.epsl.2004.10.018, 2004.

Milliman, J. D. and Syvitski, J. P. M.: Geomorphic/Tectonic Control of Sediment Discharge to the Ocean: The Importance of Small Mountainous Rivers, J. Geol., 100, 525–544, https://doi.org/10.1086/629606, 1992.

Moatar, F., Person, G., Meybeck, M., Coynel, A., Etcheber, H., and Crouzet, P.: The influence of contrasting suspended particulate matter transport regimes on the bias and precision of flux estimates, Sci. Total Environ., 370, 515–531, https://doi.org/10.1016/j.scitotenv.2006.07.029, 2006.

Montgomery, D. R. and Brandon, M. T.: Topographic controls on erosion rates in tectonically active mountain ranges, Earth Planet. Sc. Lett., 201, 481–489, https://doi.org/10.1016/S0012-821X(02)00725-2, 2002.

Mozzherin, V. V. and Sharifullin, A. G.: Estimation of current denudation rate of the mountains based on the suspended sediment runoff of the rivers (the Tien Shan, the Pamir-Alai, the Caucasus, and the Alps as an example), Geomorphol. RAS, 1, 15–23, https://doi.org/10.15356/0435-4281-2014-1-15-23, 2015.

Nash, J. E. and Sutcliffe, J. V.: River flow forecasting through conceptual models part I – A discussion of principles, J. Hydrol., 10, 282–290, https://doi.org/10.1016/0022-1694(70)90255-6, 1970.

Schliep, K. and Hechenbichler, K.: kknn: Weighted k-Nearest Neighbors, available at: https://cran.r-project.org/package=kknn (last access: 19 June 2019), 2016.

Syvitski, J. P. M. and Milliman, J. D.: Geology, geography, and humans battle for dominance over the delivery of fluvial sediment to the coastal ocean, J. Geol., 115, 1–19, 2007.

Tsyplenkov, A., Vanmaercke, M., and Golosov, V.: Caucasus-sediment-yield, available at: https://github.com/atsyplenkov/caucasus-sediment-yield (last access: 19 June 2019), 2019.

Vanmaercke, M., Poesen, J., Verstraeten, G., de Vente, J., and Ocakoglu, F.: Sediment yield in Europe: Spatial patterns and scale dependency, Geomorphology, 130, 142–161, https://doi.org/10.1016/j.geomorph.2011.03.010, 2011.

Vanmaercke, M., Poesen, J., Radoane, M., Govers, G., Ocakoglu, F., and Arabkhedri, M.: How long should we measure? An exploration of factors controlling the inter-annual variation of catchment sediment yield, J. Soils Sediment., 12, 603–619, https://doi.org/10.1007/s11368-012-0475-3, 2012.

Vanmaercke, M., Poesen, J., Govers, G., and Verstraeten, G.: Quantifying human impacts on catchment sediment yield: A continental approach, Glob. Planet. Change, 130, 22–36, https://doi.org/10.1016/j.gloplacha.2015.04.001, 2015.

Proc. IAHS, 381, 95–99, 2019
https://doi.org/10.5194/piahs-381-95-2019

Reconstruction of long-term dynamics of Chernobyl-derived ^{137}Cs in the Upa River using bottom sediments in the Scheckino reservoir and semi-empirical modelling

Alexei V. Konoplev[1], **Maxim M. Ivanov**[2], **Valentin N. Golosov**[2,3,4], **and Evgenyi A. Konstantinov**[4]

[1]Institute of Environmental Radioactivity, Fukushima University, Fukushima, 960-1296, Japan
[2]Faculty of Geography, Moscow State University, Moscow, 119991, Russia
[3]Department is Institute of Environmental Sciences, Kazan Federal University, 420008, Kazan, Russia
[4]Institute of Geography, Russian Academy of Science, Moscow, 119017, Russia

Correspondence: Alexei V. Konoplev (alexeikonoplev@gmail.com)

Published: 1 August 2019

Abstract. Two cores of bottom sediments were collected in 2018 to a depth of ~ 200 cm in the deepest part of the Scheckino reservoir on the Upa River (9500 km^2), Tula region, Russia. This area was severely contaminated by radiocesium (r-Cs) after the Chernobyl accident in 1986. The fact that ^{137}Cs activity concentrations in a specific horizon of the bottom sediments correspond to ^{137}Cs concentrations associated with suspended matter delivered to the reservoir, provides a basis for constructing the dynamics of particulate ^{137}Cs activity concentrations in the Upa River catchment from 1986 to 2017. Over the time since the Chernobyl accident, the particulate ^{137}Cs concentrations have decreased by more than an order of magnitude, with only minor changes occurring during the last 15 years. Using a typical value for the distribution coefficient K_d for the rivers of the Chernobyl contamination zone, dissolved ^{137}Cs activity concentrations in the Upa River have been estimated and their changes over the past 30 years since the accident have been studied. The resulting estimates of dissolved ^{137}Cs concentrations in the Upa River have been found to be in good agreement with measured data over the period 1987–1991. The proposed and tested method provides a basis for reconstructing the long-term dependence of radionuclide concentrations in rivers and reservoirs based on their vertical distribution in bottom sediments. Reconstructed time dependencies of particulate and dissolved ^{137}Cs activity concentrations in the Upa River were found to be described well by the proposed semi-empirical "diffusion" model based on an assumption that the time dependency of particulate r-Cs in the river corresponds to the time dependency of its concentration in top soil layers across the catchment which can be approximated by a dispersion-convection equation with physically meaningful parameters.

1 Introduction

Large areas of land were contaminated with radionuclides from the Chernobyl fallout (Izrael, 1998). These areas became long-term sources of radionuclides to natural waters and aquatic ecosystems in the Central European Russian landscape (Ivanov et al., 2017). Wash-off by surface runoff is the primary pathway for contamination of water bodies after initial fallout deposition from the atmosphere (Konoplev et al., 1992). Initial radioactive contamination of water bodies after the Chernobyl accident was relatively high as a result of direct fallout onto river and reservoir surfaces.

Over longer time periods after fallout, radionuclides bound in catchment surface soils are slowly transferred to rivers by the erosion of soil particles (particulate phase) and by desorption from soil to solution (dissolved phase). The rates of radionuclide transfer are influenced by the extent of soil erosion, the strength of radionuclide binding to catchment soils

Published by Copernicus Publications on behalf of the International Association of Hydrological Sciences.

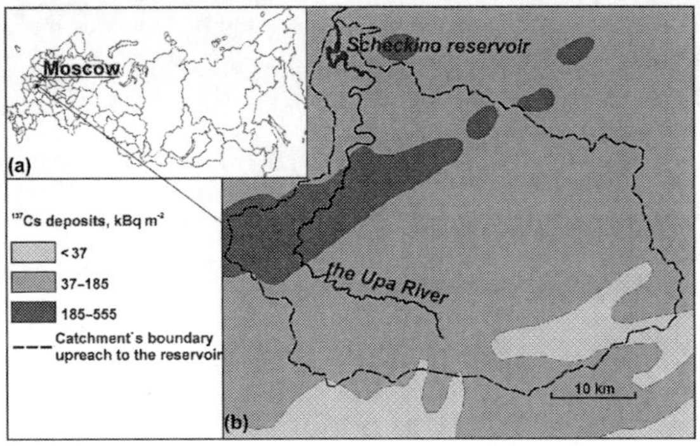

Figure 1. Location of the Upa river basin **(a)** and map of Chernobyl-derived ^{137}Cs deposition in the Upa River catchment upstream the Scheckino reservoir (Izrael, 1998) decay corrected for 1 May 1986 **(b)**.

and the potential for migration down the soil profile (Golosov et al., 2018). Detailed analysis of Chernobyl data over an extended period provides a basis for long-term prediction of changes in environmental radioactive contamination as a result of the Fukushima accident, or any other emergency involving radioactive emissions to the surrounding environment.

Unfortunately, long-term data are not always available regarding changes in radioactive contamination, and this is the case for ^{137}Cs in the Chernobyl zone rivers. Against this background, this study attempted to reconstruct, from 1986 to the present day, changes in ^{137}Cs concentrations in the Upa River as a result of the Tula-Oryol contamination zone, based on the current vertical distribution of ^{137}Cs (as of 2018) in bottom sediments of the Scheckino reservoir. These changes are described by a semi-empirical diffusion model.

2 Materials and methods

2.1 Sampling and processing of bottom sediments

The Upa River basin lies in the northern part of the Central Russian Upland within the Tula region (Fig. 1). Mean annual precipitation in the basin is $\sim 540\,\mathrm{mm\,yr^{-1}}$. The southern part of the basin, with a high proportion of arable land, was heavily contaminated by the Chernobyl fallout in April–May 1986. The northern forested part appeared to be less contaminated. Tula city is located in the north-east part of the Upa basin. The upper part of the Upa river ($1362\,\mathrm{km^2}$) is restricted by the Scheckino reservoir.

The reservoir has an area of $6\,\mathrm{km^2}$ and is elongated upwards of the dam following the drowned valley of the Upa River. For the investigation of radionuclide concentrations in bottom sediments, the upper part of the reservoir was selected since local sources of sediment such as bank erosion and inputs resulting from the transport of sediment by small streams were judged to be minor contributors.

The bottom sediments were collected in February 2018 using a pistol sampler in the deepest parts of the Scheckino reservoir in which higher rates of sedimentation were expected. The cores were wrapped in cellophane film and transported to laboratory. In laboratory, the cores were sliced into 2 cm sections, weighed, dried at 105 °C for 8 h and weighed again for determining moisture content and calculating dry sediment mass. The samples were then ground, sieved through 2 mm mesh and placed in a container of defined geometry for measurement of ^{137}Cs activity using a semi-conductor γ-spectrometer with high purity Ge detector.

2.2 Semi-empirical "diffusional" model of ^{137}Cs dynamics in river water

2.2.1 Particulate ^{137}Cs in surface runoff and rivers

Over the long term, it can be assumed that a major source of suspended matter in surface runoff and rivers is the upper layer of catchment soils. Radionuclide concentrations in top soil decrease over time due to vertical migration to deeper layers. Hence vertical migration in catchment soils is a critical process responsible for the decline in particulate-associated radionuclide concentrations in surface waters. According to the convection-diffusion model (Crank, 1975; Prokhorov, 1981; Bulgakov et al., 2002) a change in radionuclide concentration in the top soil layer C_s with time is given by the following equation:

$$C_\mathrm{s}(t) = \frac{\sigma}{\sqrt{\pi D_\mathrm{eff} t}} e^{-\left(\frac{v^2}{4D_\mathrm{eff}}+\lambda\right)t} \tag{1}$$

where D_eff is the effective dispersion coefficient; v is the effective velocity of convective transport; λ is the radioactive decay constant; σ is the radionuclide deposition density, and t is time.

The time dependence of particulate radionuclide concentrations C_p in surface runoff and river water can be described by this equation using averaged values of D_eff and v over the catchment area.

Studies of r-Cs vertical migration in soils of river catchments show that, as a rule, its transport due to dispersion prevails over its convective transport (Konshin, 1992; Ivanov et al., 1997; Konoplev et al., 2016). Therefore,

$$\frac{v^2}{4D_\mathrm{eff}} \ll \lambda \tag{2}$$

Eq. (1) can be simplified as follows:

$$C_\mathrm{p}(t) = \frac{\sigma}{\sqrt{\pi D_\mathrm{eff} t}} e^{-\lambda t} \sim \frac{e^{-\lambda t}}{\sqrt{t}} \tag{3}$$

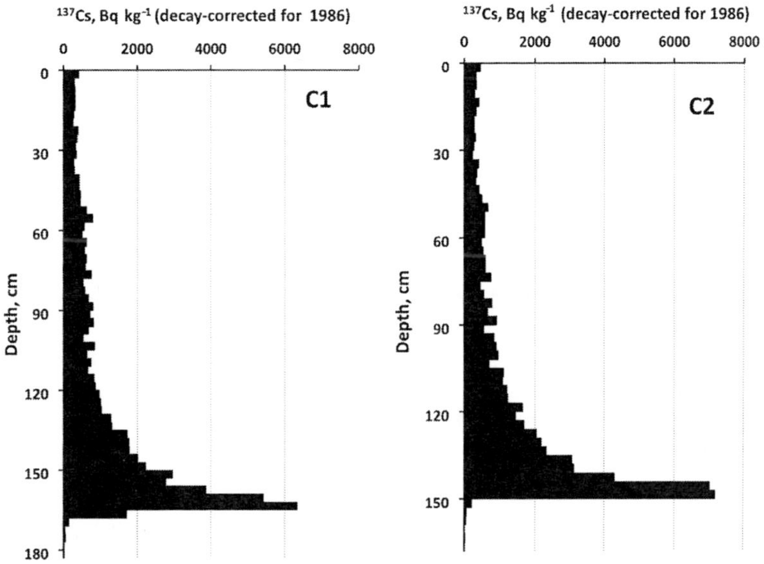

Figure 2. Vertical distribution of ^{137}Cs, decay-corrected for 1986, in the reservoir cores. The total inventory of ^{137}Cs in core C1 is 1030 kBq m^{-2} compared with 1160 kBq m^{-2} in core C2.

2.2.2 Dissolved ^{137}Cs in surface runoff and rivers

Dissolved radionuclides in surface runoff and rivers are transferred from soil to water due to cation exchange. As this takes place, only the exchangeable fraction of radionuclides is involved in exchanges with the dissolved phase. Therefore, dissolved radionuclide concentrations in water C_d relate to particulate concentrations as follows (Konoplev and Bulgakov, 2000):

$$C_\mathrm{d}(t) = \frac{C_\mathrm{p}(t) \cdot \alpha_\mathrm{ex}}{K_\mathrm{d}^\mathrm{ex}} = \frac{C_\mathrm{p}(t)}{K_\mathrm{d}^\mathrm{tot}} \tag{4}$$

where α_ex is the exchangeable fraction of the radionuclide, and $K_\mathrm{d}^\mathrm{tot}$ and K_d^ex are, respectively, the total and exchangeable distribution coefficients of the radionuclide.

Over longer time periods, when equilibrium between exchangeable and nonexchangeable forms of r-Cs is reached and when condition (2) is valid, the equation for radionuclide dissolved concentrations can be presented as follows:

$$C_\mathrm{d}(t) = \frac{\alpha_\mathrm{ex}(\infty)\sigma}{K_\mathrm{d}^\mathrm{ex}\sqrt{\pi D_\mathrm{eff} t}} e^{-\lambda t}$$

$$= \frac{\sigma}{K_\mathrm{d}^\mathrm{tot}\sqrt{\pi D_\mathrm{eff} t}} e^{-\lambda t} \sim \frac{e^{-\lambda t}}{\sqrt{t}} \tag{5}$$

The advantage of this semi-empirical "diffusion-based approach" is that all phases after the accident can be described by the same equation using the same physically-based parameters which can be estimated or determined in the field or using laboratory studies. Generally speaking, the decay-corrected particulate and dissolved ^{137}Cs activity concentrations in both surface runoff and rivers should follow the inverse square root of the time function.

3 Results and discussion

Figure 2 presents the vertical distribution of ^{137}Cs activity concentrations in the two bottom sediment cores collected from the study reservoir (C1 and C2). Accumulation of sediments entering the reservoir is associated with an increase in the total inventory of ^{137}Cs. In both cores, the ^{137}Cs inventory is more than twice the corresponding maximum inventory for the Upa River basin (Fig. 2)

Regretfully, no systematic long-term monitoring of radioactive contamination has been conducted on the Upa River. Data are only available for the first few years after the accident (1987–1991) for total activity concentrations of ^{137}Cs in water (dissolved + particulate) on the Upa River (Vakulovsky et al., 1994). The data obtained by the present study regarding the vertical distribution of ^{137}Cs in bottom sediments in deep-water zones, where sediments tend to accumulate, can provide a basis for reconstructing temporal changes in both particulate and dissolved ^{137}Cs concentrations in the Upa River water following the Chernobyl accident. Here, we make an assumption that annual sediment accumulation on the bottom of the reservoir was not changing over the years since the accident. The shape of the ^{137}Cs vertical distribution in the cored sediment accumulation zones

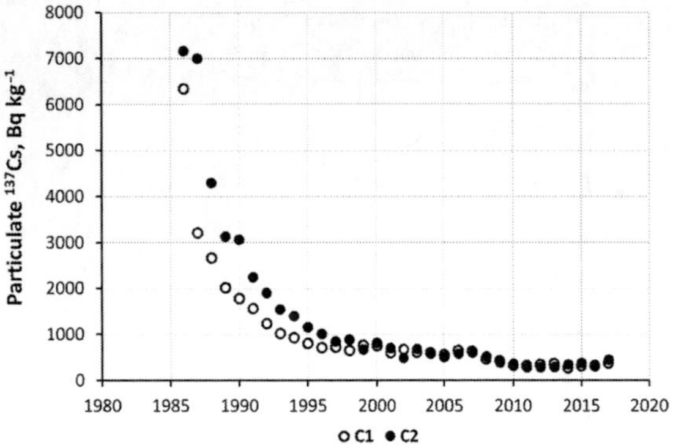

Figure 3. Reconstructed time dependencies of particulate ^{137}Cs activity concentrations (decay-corrected for 1986) in the Upa River based on ^{137}Cs vertical distributions in cores C1 and C2 collected in February 2018.

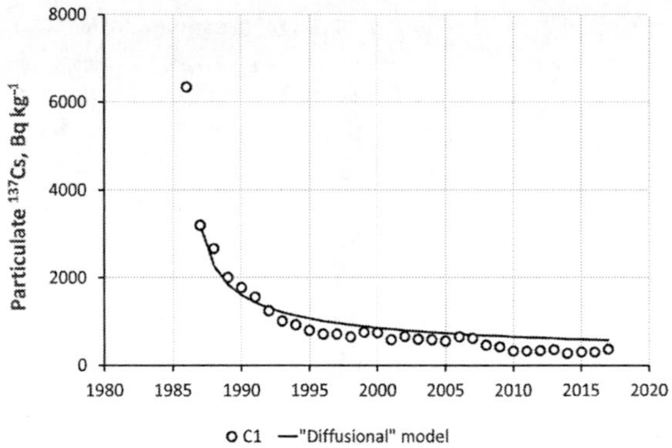

Figure 4. Diffusion based semi-empirical modelling of reconstructed particulate ^{137}Cs in the Upa River (line) based on the depth profile for the bottom sediments in core C1 (points).

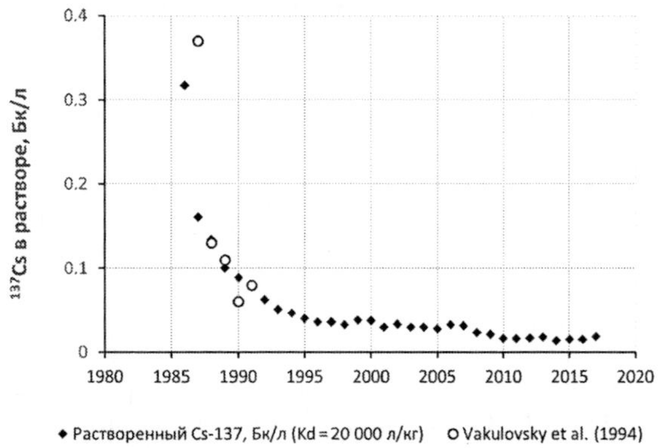

Figure 5. Comparison of the predicted time dependence of dissolved ^{137}Cs activity concentrations in the Upa River based on reconstructed particulate ^{137}Cs from sediments in core C1 with published measurements for 1987–1991 (Vakulovsky et al., 1994). To calculate dissolved ^{137}Cs activity concentrations, the total distribution coefficient $K_d^{tot} = 20\,000\,\mathrm{L\,kg^{-1}}$ (Konoplev, 2015) was used.

(Fig. 2) indicates that depositing sediments do not mix vertically and that the ^{137}Cs incorporated in them does not migrate, since the ^{137}Cs peak is well defined. Considering the mean rate of sediment accumulation is about 5 cm yr^{-1}, based on the position of the peak radionuclide concentration (Fig. 2), the bottom sediment layers can be attributed to a certain period of sediment accumulation. If ^{137}Cs activity concentration in a respective layer matches its concentration on suspended matter in this time period, changes in ^{137}Cs concentrations on suspended matter in the Upa River from 1986 to 2017 can be easily derived.

Figure 3 shows a reconstructed time dependence of particulate ^{137}Cs concentration in the Upa river, based on the data for the two bottom sediment cores collected in February 2018. As can be seen, over the time since the accident, ^{137}Cs concentrations have decreased by more than an order of magnitude, with only minor changes during the last 15 years.

Figure 4 shows the results of calculating temporal changes in ^{137}Cs concentrations in suspended material in the Upa River by the described model (line) with concentrations reconstructed based on the sediment core C1. As can be seen, the "diffusion" model provides a good representation of the time dependency of particulate ^{137}Cs concentrations in the Upa River reconstructed using the depth profile in the bottom sediments.

Further, using the typical value of the distribution coefficient K_d for the rivers of the Chernobyl contaminated zone, it is possible to estimate the concentration of dissolved ^{137}Cs in the Upa River and its variation over the 30 years following the accident. A K_d of 20 000 L kg^{-1} was taken as being a representative value on the basis of analysis of data for a number of rivers in Chernobyl contaminated areas (Konoplev, 2015). Using this value, and reconstructed data for particulate ^{137}Cs activity concentrations, it is possible to estimate the concentrations of dissolved ^{137}Cs in the Upa River for the time window from 1986 to 2017. The results of these estimations are presented in Fig. 5. The published measurement data for 1987–1991 (Vakulovsky et al., 1994) is also given for comparison. It can be seen that the predicted activity concentrations of dissolved ^{137}Cs in the Upa River and their temporal trend are consistent with the measured data.

Data availability. The data can be accessed by request to authors.

Author contributions. AVK suggested the idea and semi-empirical model, MMI, VNG and EAK designed the sampling and provided radiocesium analysis, AVK analyzed the data and prepared the manuscript with contributions from all co-authors.

Competing interests. The authors declare that they have no conflict of interest.

Special issue statement. This article is part of the special issue "Land use and climate change impacts on erosion and sediment transport". It is a result of the ICCE Symposium 2018 – Climate Change Impacts on Sediment Dynamics: Measurement, Modelling and Management, Moscow, Russia, 27–31 August 2018.

Financial support. This research has been supported by Russian Foundation for Basic Research (RFBR project no. 18-55-50002) and the Japan Society for the Promotion of Science (JSPS) within the framework of the bilateral project "Assessment and prediction of sediment and radionuclide fluxes in a river basin affected by a severe accident".

References

Bulgakov, A. A., Konoplev A. V., Kanivets, V. V., and Voitsekhovich, O. V.: Modelling the long-term dynamics of radionuclides in rivers, Radioprotection – Colloques, 37, 649–654, 2002.

Crank, J.: The mathematics of diffusion, Oxford University Press, 56–61, London, UK, 1975.

Golosov, V. N., Walling, D. E., Konoplev, A. V., Ivanov, M. M., and Sharifullin, A. G.: Application of bomb- and Chernobyl-derived radiocaesium for reconstructing changes in erosion rates and sediment fluxes from croplands in areas of European Russia with different levels of Chernobyl fallout, J. Environ. Radioactiv., 186, 78–89, https://doi.org/10.1016/j.jenvrad.2017.06.022, 2018.

Ivanov, M. M., Golosov, V. N., and Belyaev, V. R.: Analysis of topography structure for evaluation of sediment delivery ratio within the Plava river basin (Tula oblast), Moscow University Bulletin, ser. 5, Geography, 3, 14–23, 2017.

Ivanov, Y. A., Lewyckyj, N., Levchuk, S. E., Prister, B. S., Firsakova, S. K., Arkhipov, N. P., Kruglov, S. V., Alexakhin, R. M., Sandalls, J., and Askbrant, S.: Migration of ^{137}Cs and ^{90}Sr from Chernobyl fallout in Ukrainian, Belarussian and Russian soils, J. Environ. Radioactiv., 35, 1–21, 1997.

Izrael, Y. A. (Ed.): Atlas of radioactive contamination of European part of Russia, Belarus and Ukraine, Roshydromet, Roskartography, Moscow, Russia, 142 pp., 1998 (in Russian).

Konoplev, A. V.: Distribution of radiocesium of accidental origin between the suspended matter and solution in rivers: Comparison of Fukushima and Chernobyl, Radiochemistry, 27, 552–556, 2015.

Konoplev, A. V. and Bulgakov, A. A.: Exchangeable distribution coefficient of ^{90}Sr and ^{137}Cs in soil-water system, Atom. Energy+, 88, 158–163, 2000.

Konoplev, A. V., Bulgakov, A. A., Popov, V. E., and Bobovnikova, T. I.: Behaviour of long-lived Chernobyl radionuclides in a soil–water system, Analyst, 117, 1041–1047, 1992.

Konoplev, A. V., Golosov, V., Laptev, G., Nanba, K., Onda, Y., Takase, T., Wakiyama, Y., and Yoshimura, K.: Behavior of accidentally released radiocesium in soil-water environment: looking at Fukushima from a Chernobyl perspective, J. Environ. Radioactiv., 151, 568–578, https://doi.org/10.1016/j.jenvrad.2015.06.019, 2016.

Konshin, O. V.: Mathematical model of ^{137}Cs migration in soil: analysis of observations following the Chernobyl accident, Health Phys., 63, 301–306, 1992.

Prokhorov, V. M.: Migration of radioactive contaminants in soils: physico-chemical mechanisms and modelling, Energoatomizdat, Moscow, Russia, 99 pp., 1981 (in Russian).

Vakulovsky, S. M., Nikitin, A. I., Chumichev, V. B., Katrich, I. Y., Voitsekhovich, O. V., Medinets, V. I., Pisarev, V. V., Bovkun, L. A., and Khersonsky, E. S.: Cesium-137 and strontium-90 contamination of water bodies in the areas affected by releases from the Chernobyl nuclear power plant accident: an overview, J. Environ. Radioactiv., 23, 103–122, 1994.

Proc. IAHS, 381, 101–106, 2019
https://doi.org/10.5194/piahs-381-101-2019

Temporal trends of ^{137}Cs activity concentration in pond waters in the vicinity of Fukushima Dai-ichi nuclear power plant

Yoshifumi Wakiyama[1], **Alexei Konoplev**[1], **Toshihiro Wada**[1], **Tsugiko Takase**[1], **Yasunori Igarashi**[1], **Kenji Nanba**[1,2], **and Ian Byrnes**[3]

[1]Institute of Environmental Radioactivity, Fukushima University, Fukushima, 960-1296, Japan
[2]Faculty of Symbiotic Systems Science, Fukushima University, Fukushima, 960-1296, Japan
[3]Faculty of Environmental Sciences and Natural Resource Management, Norwegian University of Life Science, Aas, 1430, Norway

Correspondence: Yoshifumi Wakiyama (wakiyama@ipc.fukushima-u.ac.jp)

Published: 1 August 2019

Abstract. Closed and semi-closed water bodies, such as lakes and ponds, are important water resources in Fukushima area and they are the most sensitive environments to radioactive contamination after the Fukushima Dai-ichi nuclear power plant accident. Wakiyama et al. (2017) investigated ^{137}Cs activity in water and bottom sediment in four ponds; Suzuuchi (SU), Funasawa (FS), Inkyozaka (IZ), and Kashiramori (KM), within 10 km zone from the FDNPP during 2015–2016. This study follows up their observation to address longer time trends of ^{137}Cs activity concentration in pond waters and to show speciation of ^{137}Cs in soil and bottom sediment. Mean total ^{137}Cs activity concentration in water ranged from 2.5 to 29 Bq L^{-1}. There was not found steady trend in the activity concentration of total and particulate ^{137}Cs for four ponds. The concentration of dissolved ^{137}Cs was usually low in winter during the entire observation period in all four ponds. A tendency to a decrease in the ^{137}Cs activity concentration in suspended sediments was found for four ponds and the decreasing rate constants, including radiological decay, on SU, FS, IZ and KM were 0.33, 0.53, 0.29 and 0.25 yr^{-1}, respectively. The results of sequential extractions of soil and bottom sediment samples showed higher proportion of bioavailable ^{137}Cs, i.e., exchangeable and organic bound ^{137}Cs, in bottom sediment than in the soil.

1 Introduction

Closed and semi-closed water bodies, such as lakes and ponds, are the most sensitive environments to radionuclide contamination. More than 3700 individual irrigation ponds of different sizes are located in Fukushima Prefecture, which are radiologically contaminated after the Fukushima Dai-ichi nuclear power plant (FDNPP) accident. The Tohoku Regional Agricultural Administration Office of the Ministry of Agriculture, Forestry and Fishery (MAFF) and the Fukushima Prefectural government (MAFF and Fukushima prefecture, 2015) investigated 2679 ponds in Fukushima Prefecture and reported their water and sediment contamination levels.

The long-term behaviour of radiocesium in a closed and semi-enclosed water system affected by the Chernobyl accident can serve as a basis for long-term prediction of changes in environmental radioactive contamination in Fukushima. For example, Putyrskaya et al. (2009) presented the results of more than ten years of observation on the behaviour of ^{137}Cs in several lakes affected by the Chernobyl accident and elaborated a model calculation of temporal variations in the concentration of ^{137}Cs activity concentration of the lake water. However, the meteorological and geographical conditions in these areas are different from those in the Fukushima area, and the applicability of these results requires verification (Konoplev et al., 2016).

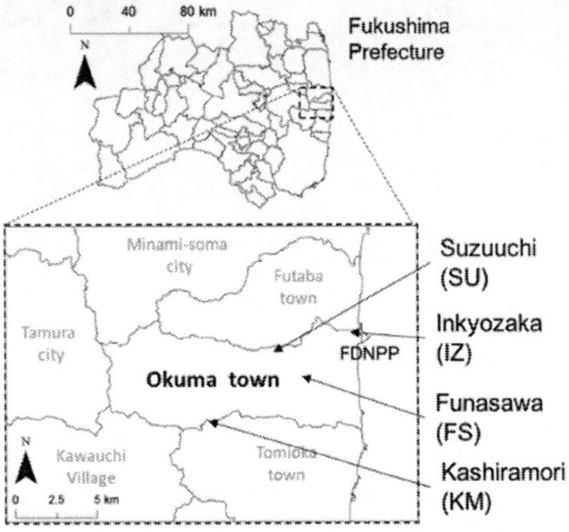

Figure 1. Location of study sites in Fukushima Prefecture and in Okuma town.

Wakiyama et al. (2017) demonstrated radiological contaminations in four ponds in the vicinity of FDNPP based on one-year observation. Although they suggested seasonal variation of dissolved ^{137}Cs activity concentration of water, long-term trends remained unclear due to short period of their observations. This study shows longer temporal trends of ^{137}Cs activity concentration in pond waters, as a continuation of observations undertaken by Wakiyama et al. (2017). In addition, we evaluated contamination of bottom sediment and its speciation.

2 Materials and method

2.1 Study site

Selected irrigation ponds of Suzuuchi (SU), Funasawa (FS), Inkyozaka (IZ) and Kashiramori (KM) are located in the city of Okuma within a radius of 10 km from FDNPP (Fig. 1). The surface area of these ponds of SU, FS, IZ and KM was 4100, 10 700, 6500 and 8100 m^2, respectively, and ambient dose equivalent rates measured on March 2014 around SU, FS, IZ and KM were 24, 6.0, 11 and 3.3 µSv h^{-1}, respectively (Wakiyama et al., 2017). The average annual precipitation and the average annual air temperature in this region were 1493 mm and 12.3 °C, respectively, according to Namie weather station data from 1976 to 2015.

2.2 Sampling and analyses of pond water

We had collected 2 L of water samples from the four ponds approximately once a month since August 2015 to August 2017. Then, water samples were filtered using 0.45 µm-pore cellulose acetate membrane filter units in the laboratory. Suspended sediment (SS) concentrations in water (SSC, kg L^{-1}) were quantified by dividing weight of SS by volume of filtered water. Suspended sediments and filtrates were measured with Ge detectors to quantify the ^{137}Cs activity concentration in SS ($C_{\text{Cs-SS}}$, Bq kg^{-1}) and dissolved ^{137}Cs activity concentration ($C_{\text{Cs-dis}}$, Bq L^{-1}), respectively. Measurements of ^{137}Cs activity were done with standard electrode coaxial Ge detectors (CANBERRA GC4020, Canberra, USA) with relative efficiency of 42.6 % at the Institute of Environmental Radioactivity, Fukushima University. Measurements were taken for 90 000 s for ^{137}Cs activity concentrations in SSs and 240 000 s for dissolved ^{137}Cs activity concentrations to ensure minimal statistical errors, which were < 5 % and < 10 %, respectively.

Particulate ^{137}Cs activity concentration in water ($C_{\text{Cs-par}}$, Bq L^{-1}), total ^{137}Cs activity concentration in water ($C_{\text{Cs-tot}}$, Bq L^{-1}), and the apparent distribution coefficient (K_d, L kg^{-1}), were calculated by following equations.

$$C_{\text{Cs-par}} = C_{\text{Cs-SS}} \times \text{SSC} \quad (1)$$

$$C_{\text{Cs-tot}} = C_{\text{Cs-par}} + C_{\text{Cs-dis}} \quad (2)$$

$$K_d = C_{\text{Cs-SS}} \div C_{\text{Cs-dis}} \quad (3)$$

2.3 Sampling and analyses of soil and bottom sediment

Soil and bottom sediment samples were collected in June 2016. Soils at a depth of 30 cm using a liner core sampler DIK-110C (DAIKI, Japan) with a plastic cylinder insert 5 cm in diameter. In laboratory, surface soils, to the depths of 3–5 cm, were cut for every 1–2 cm, and then remained subsurface soils were cut by 3–5 cm. Each soil sample was dried at 50 °C for at least three days, then ground and homogenized in a mortar. Three cores of bottom sediment were obtained using 4 cm diameter polyvinyl core tubing in each pond. The bottom sediment cores were taken from a boat and the sampling depth ranged from 33 to 56 cm. Bottom sediments to the depth of 10 cm were cut for interval of 2.5–5 cm, and subsurface bottom sediments were cut by 5–10 cm.

Each bottom sediment sample was dried at 50 °C for at least three days, then ground and homogenized in a mortar. The homogenised samples of both soil and bottom sediments were packed into 100 mL containers and measured to determine its ^{137}Cs activity concentration (Bq kg^{-1}). The measurements were done with standard electrode coaxial Ge detectors (CANBERRA GC4020, Canberra, USA) with relative efficiency of 42.6 % at the Institute of Environmental Radioactivity, Fukushima University. Measurements were taken up to 85 600 s to ensure a minimal statistical error of < 5 % for almost samples. Eight samples taken from the deep portion of the bottom sediment had low ^{137}Cs activity concentrations, at a minimum of 1.83 Bq kg^{-1}, and high relative error, at 12.8 % maximum.

Table 1. Summary of ^{137}Cs activity concentration in pond waters.

	SU			FS			IZ			KM		
	n	Mean	S.D*	*n*	Mean	SD*	*n*	Mean	S.D*	*n*	Mean	SD*
Dissolved ^{137}Cs activity concentration ($Bq\,L^{-1}$)	24	2.5	1.4	21	1.7	0.76	24	1.9	0.64	12	0.25	0.085
Particulate ^{137}Cs activity concentration ($Bq\,L^{-1}$)	24	26	29	24	9.5	11	24	7.2	9.1	24	1.3	1.5
Total ^{137}Cs activity concentration ($Bq\,L^{-1}$)	24	30	22	21	12	7.2	24	9.2	3.8	12	1.5	1.1
^{137}Cs activity concentration in SS ($kBq\,kg^{-1}$)	24	290	95	24	270	120	24	250	110	24	130	86
Apparent distribution coefficient, K_d ($10^5\,L\,kg^{-1}$)	24	1.4	0.45	21	2.0	0.93	24	1.4	0.75	12	6.8	6.2

* SD means standard deviation (1σ).

The proportion of ^{137}Cs in the exchangeable fraction, the organic fraction and the fraction strongly bound to the particles in the soil and bottom sediment were evaluated by a sequential extraction. Soil samples for quantifying the proportion were prepared by mixing up 0–1 and 1–2 cm of soil samples which are described in Sects. 2–3. The soil samples were passed through 2 mm-mesh sieves. Bottom sediment for quantifying the proportion were additionally taken by an Ekman-Berge sampler on June 2016, simultaneous to bottom the core sampling. The bottom samples were dried at 50 °C.

The exchangeable fraction of the soil and bottom sediment were extracted with 1 M CH_3COONH_4 and the extract was measured for ^{137}Cs activity. Subsequently, organic matter was decomposed with mixture of 5 % of H_2O_2 solution and nitric acid. The extract was measured for ^{137}Cs activity. The measurements were done with same Ge detectors for soil and bottom sediment samples. Measurements were taken up to 53 000 s to ensure a minimal statistical error of < 5 %. Only for the organic-bound fraction of ^{137}Cs in bottom sediment from IZ, relative error was 14.6 % because of low ^{137}Cs activity.

3 Results and discussion

3.1 ^{137}Cs activity concentrations in pond water

The observation results are presented in Table 1. Particulate ^{137}Cs accounted for 73 %–88 % of total ^{137}Cs, on average. The mean apparent K_d value ranged from 1.4×10^5 to $6.0 \times 10^5\,L\,kg^{-1}$. These values were higher than those observed in the Chernobyl area by at least one order of magnitude, as previous authors suggested (e.g., Tsukada and Ohse, 2016). These results show a stronger affinity of ^{137}Cs for sediment particles in Fukushima.

Figure 2 shows temporal variations in total, particulate and dissolved ^{137}Cs activity concentration in pond water. Both total and particulate ^{137}Cs activity concentrations varied significantly throughout observation, and the temporal variations were similar. These similar variations were attributed to high proportions of particulate ^{137}Cs activity concentration, as shown in Table 1. In contrast, the concentration of dissolved ^{137}Cs was usually low in winter during the entire observation period in all ponds. These results confirm that the seasonal cycle exists on dissolved ^{137}Cs in ponds, as it was previously found in the Chernobyl-affected area (e.g., Putyrskaya et al., 2009).

Decreasing trends of ^{137}Cs activity concentration in SSs were found for all ponds (Fig. 3). The temporal trend of ^{137}Cs activity concentration in SS was expressed by the following equation,

$$C_{\text{Cs-SS}}(t) = C_{\text{Cs-SS},0}\exp(-\lambda t) \quad (4)$$

where t is the elapsed time since the accident (year), $C_{\text{Cs-SS}}(t)$ is ^{137}Cs activity concentration in SSs at time t ($Bq\,kg^{-1}$), $C_{Cs-SS,0}$ is initial ^{137}Cs activity concentration in SSs at $t = 0$ ($Bq\,kg^{-1}$) and λ is rate constant including radiological decay (yr^{-1}). The rate constant on SU, FS, IZ and KM were 0.33, 0.52, 0.29 and 0.25 yr^{-1}, respectively. The corresponding periods of half-reduction for SU, FS, IZ

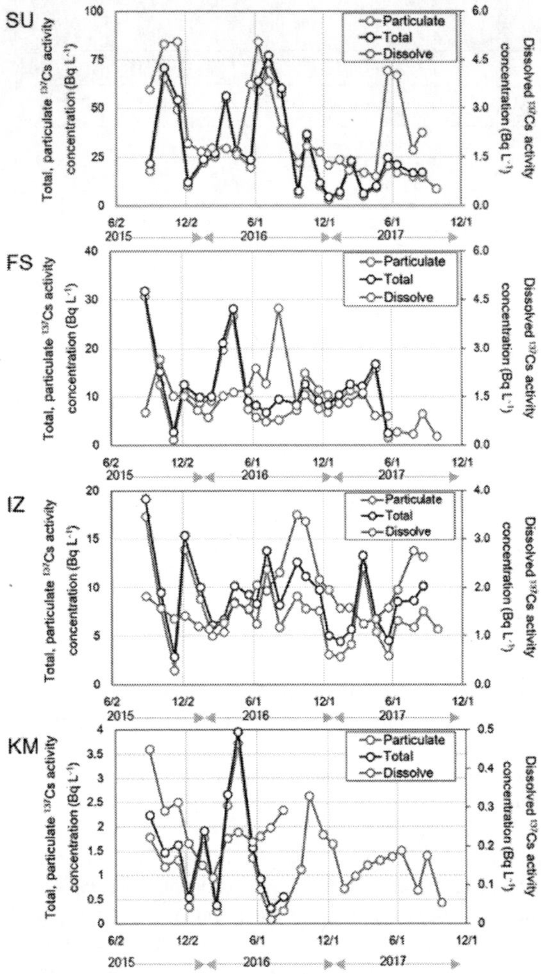

Figure 2. Temporal variations in total, particulate, dissolved ^{137}Cs activity concentration in the different ponds.

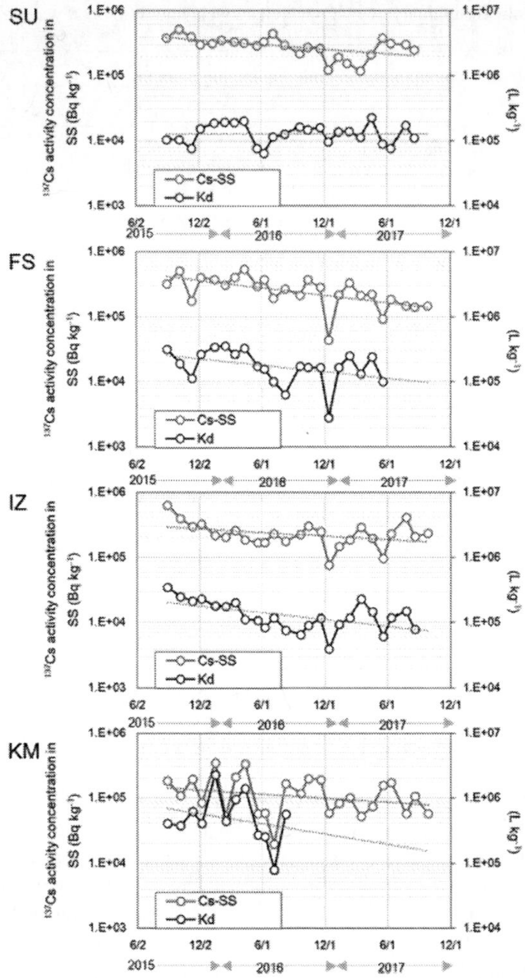

Figure 3. Temporal variations in ^{137}Cs activity concentration of suspended sediment (Cs-SS) and apparent distribution coefficient K_d in the different ponds.

and KM were 2.1, 1.3, 2.4 and 2.8 years, respectively. Since ^{137}Cs activity concentration in SS was more variable than that dissolved ^{137}Cs activity concentration, the temporal variation in apparent K_d seemed to coincide with changes in the ^{137}Cs activity concentration in SS.

3.2 ^{137}Cs in soil and bottom sediment

The vertical distributions of ^{137}Cs activity concentration in soil and bottom sediment are relatively similar for the all study sites (Fig. 4). As indicated by Wakiyama et al. (2017), the mean ^{137}Cs inventory in the soil around SU, FS, IZ and KM were 6.8, 4.6, 2.9 and 0.42 MBq m^{-2}, respectively. Similarly, mean ^{137}Cs inventory in bottom sediment in SU, FS, IZ and KM were 13, 8.9, 1.6 and 1.1 MBq m^{-2}, respectively. The ^{137}Cs activity concentration both in the soils and bottom sediments decreased with depth.

The depth at which the soil contains 90 % of the inventory of ^{137}Cs in the soil around the SU, FS, IZ and KM were 5, 7.6, 9.7 and 6.7 cm, respectively. The depth at which the soil contains 90 % of the inventory of ^{137}Cs in bottom sediment on SU, FS, IZ and KM were 30, 17, 8.3 and 12.5 cm, respectively. Deeper penetration of ^{137}Cs into bottom sediments than in the soil were found with the exception of IZ.

At least 95 % of ^{137}Cs in soil was strongly bound to sediment particle in soil, while the ^{137}Cs content in bottom sediment ranged from 42 % to 72 % (Fig. 5). The low proportions of ^{137}Cs strongly bound with particle in bottom sedi-

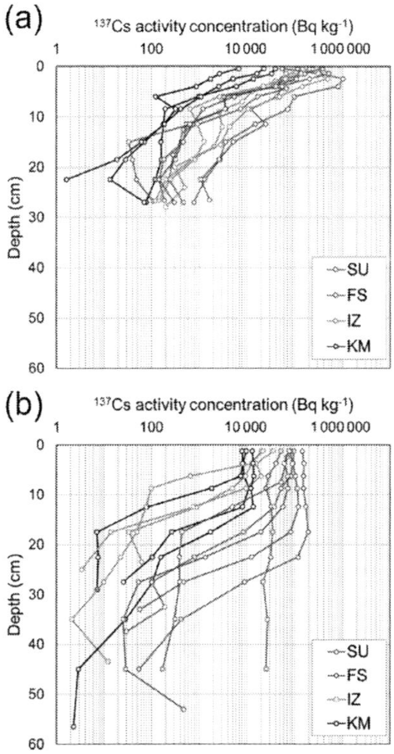

Figure 4. Vertical distribution of ^{137}Cs activity concentration in soil (**a**) and bottom sediment (**b**).

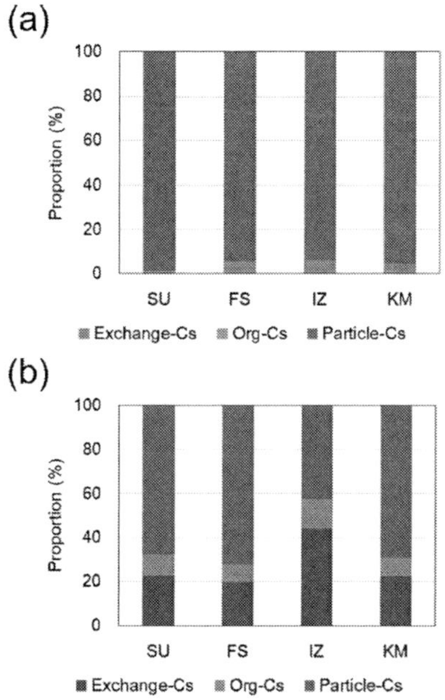

Figure 5. Proportion of exchangeable, organic and particle bound fractions of ^{137}Cs in soil of 0–2 cm depth (**a**) and bottom sediment by an Ekman–Berge sampler (**b**).

ment, suggest that the bioavailability of ^{137}Cs was increased in bottom sediments.

Putyrskaya et al. (2009) demonstrated that percentage of exchangeable ^{137}Cs was 5.7 % in average in 2006, based on sequential extraction of bottom sediment taken in lakes affected during the Chernobyl accident. Although the depth and water flow regime of ponds, located near Chernobyl, differed from ours, a high proportion of ^{137}Cs in the exchangeable and organic fractions in Fukushima study sites suggested that the bioavailability of ^{137}Cs in Okuma town ponds is still high.

These mobilities may have led to the vertical migration of ^{137}Cs and controlled the fluctuation in dissolved ^{137}Cs activity concentration. The effect of biological activities on ^{137}Cs behaviour should be high in the spotlight.

4 Summary

We investigated the behaviour of ^{137}Cs in four small ponds around Fukushima Dai-ichi nuclear power plant as a following-up of observation by Wakiyama et al. (2017). Continuous monitoring of ^{137}Cs activity concentration in water confirmed the presence of a seasonal cycle of dissolved ^{137}Cs activity concentration and newly revealed tendencies to a decrease trends in ^{137}Cs activity concentration in suspended sediment for all ponds.

In addition, the sequential extractions of sediment samples showed a higher proportion of ^{137}Cs in the exchangeable and organic fractions in bottom sediments than in soils and this indicates an increase in the bioavailability of ^{137}Cs in bottom sediment. The effect of biological activity should be tested to better understand ^{137}Cs activity in ponds.

Data availability. The data can be accessed by request to authors.

Author contributions. YW: Sampling, analyses and discussion as corresponding author. AK: Sampling, sample treatment and discussion. TW: Sampling, and discussion. TT: Sample analyses, and discussion. YI: Sampling, analyses, and discussion. KN: Sampling, discussion, and facilitating observation. IB: Sampling, sample treatment, and discussion.

Competing interests. The authors declare that they have no conflict of interest.

Special issue statement. This article is part of the special issue "Land use and climate change impacts on erosion and sediment transport"" It is a result of the ICCE Symposium 2018 – Climate Change Impacts on Sediment Dynamics: Measurement, Modelling and Management, Moscow, Russia, 27–31 August 2018.

Acknowledgements. This study was supported by a grant-in-aid from the Kurita Water and Environment Foundation (No. 15B098 and 16K023). This study was partly supported by JSPS KAKENHI project 18H03389. The authors would like to thank Kenji Watanabe, Azusa Goto and Jun Obara in Institute of Environmental Radioactivity, Fukushima University for supporting sampling and measurements in this study.

Financial support. This research has been supported by the Kurita Water and Environment Foundation (grant no. 15B098 and 16K023) and partly supported by JSPS KAKENHI (project 18H03389).

References

Konoplev, A., Golosov, V., Laptev, G., Nanba, K., Onda, Y., Takase, T., Wakiyama, Y., and Yoshimura, K.: Behavior of accidentally released radiocesium in soil-water environment: looking at Fukushima from a Chernobyl perspective, J. Environ. Radioact., 151, 568–578, https://doi.org/10.1016/j.jenvrad.2015.06.019, 2016.

MAFF and Fukushima prefecture: About actual situation of radionuclides in ponds etc: an overview of investigation results obtained by a demonstration project for countermeasures against expanding contamination in ponds etc., available at: http://www.maff.go.jp/j/kanbo/joho/saigai/pdf/tameike_tyousa26.pdf (last access: 8 December 2016), 2015.

Putyrskaya, V., Klemt, E., and Röllin, S.: Migration of ^{137}Cs in tributaries, lake water and sediment of Lago Maggiore (Italy, Switzerland) – analysis and comparison with Lago di Lugano and other lakes, J. Environ. Radioact., 100, 35–48, 2009.

Tsukada, H. and Ohse, K.: Concentration of radiocaesium in rice and irrigation water, and soil management Practices in Oguni, Date, Fukushima, Integr. Environ. Assess. Manage., 12, 659–661, 2016.

Wakiyama, Y., Konoplev, A., Wada, T., Takase, T., Byrnes, I., Carradine, M., and Nanba, K.: Behavior of ^{137}Cs in ponds in the vicinity of the Fukushima Dai-ichi Nuclear Power Plant, J. Environ. Radioact., 178–179, 367–376, 2017.

Proc. IAHS, 381, 107–112, 2019
https://doi.org/10.5194/piahs-381-107-2019

Projections of Selenga river runoff in the XXI century and uncertainty estimates

Tatiana Millionshchikova

Water Problems Institute of the Russian Academy of Sciences, Moscow, 119333, Russia

Correspondence: Tatiana Millionshchikova (tatyana.million@mail.ru)

Published: 1 August 2019

Abstract. The Selenga River, which originates in Mongolia, contributes nearly 50 % of the total inflow into Lake Baikal. Since 1996, the longest low-flow period has been observed. This paper focuses on the spatially distributed process-based modeling to assess possible runoff changes under climate projections in the XXI century using an ensemble of global climate models (GCMs) from ISI-MIP2 (Inter-Sectoral Impact Model Intercomparison Project, phase 2) and RCP-scenarios (Representative Concentration Pathway) as inputs. The ECOMAG hydrological model was applied to simulate possible runoff changes in the Selenga River basin. According to the simulations, the low-flow runoff into Lake Baikal will be continued throughout the XXI century under all the RCP-scenarios. Furthermore, as hydrological projections for the future are characterized by significant climate projections uncertainty, ANOVA (analyses of variance) test was used to quantify the sources of this uncertainty. The calculations showed that the contribution of the model uncertainty (differences between GCMs) is much higher than the contribution of the scenario uncertainty (variability of RCP-scenarios) although it increases at the end of the century.

1 Introduction

Modern spatially distributed process-based hydrological models are increasingly applying for the analysis of the mechanisms of hydrological system responses to future climate changes. The constructed scenarios of future climate can be simulated by climate models and then used as inputs into regional hydrological models (e.g. Gelfan et al., 2015). Implementation of such models is among key problems in catchment hydrology.

The transboundary Selenga River is an important area for hydrological research a being important water and contaminant sources for Lake Baikal (Chalov et al., 2015). Location in two countries makes a process of development of a distributed process-based hydrological model rather difficult. This is due to the sparseness of observation data and the lack of information about the characteristics of the underlying surface. Here, since 1996, a significant decrease in annual runoff amount has been observed (Frolova et al., 2017), which is caused by regional climate changes (Törnqvist et al., 2014). Due to uncertainty in estimated projections inflow to Lake Baikal, the continued increase in hydro-climatic and anthropogenic impacts creates additional difficulties for water resources management in the region. Therefore, the problem of developing a robust hydrological model for the Selenga River basin has become especially urgent.

Most of the studies of the impact of current and future climate changes on the water regime in Selenga River basin are described only for the Mongolian part of the basin (Batima, 2006; Menzel et al., 2008; Malsy et al., 2013; Hülsmann et al., 2015). There is hydro-climatic change assessment for the entire Selenga River basin using global hydrological model (Törnqvist et al., 2014), but calculations based on global models contain considerable uncertainty (Krysanova et al., 2018), and their applicability is not obvious. The regional hydrological model for the entire Selenga River basin is presented in two recent publications (Moreido and Kalugin, 2017; Karthe et al., 2017), but the results do not allow to assess the model ability to reproduce the decreasing trend in long-term variations of observed runoff.

In last decade, there is a better agreement between the results of calculations obtained by global climate models (GCMs) and the ones obtained by regional hydrological models with higher resolutions (Flato et al., 2013). However, even if the GCM-based hydrological model reproduces historical hydrological data correctly, the hydrological projections in the XXI century are uncertain (Gelfan et al., 2017). One main outcome of the recent studies states that the uncertainty is caused primarily by GCMs, rather than by the hydrological modeling (see review in Joseph et al., 2018).

The thorough analysis of the main sources of uncertainty in the entire model and scenario chain is considered in a significantly smaller number of works. Recently, more studies describing the application of the analysis of variance (ANOVA) test for allocation of uncertainty sources in climate impact analysis have appeared (see, e.g. in Hattermann et al., 2018). It was shown that the largest contribution in uncertainty is related to GCMs, followed by Representative Concentration Pathway (RCP) scenarios, and the smallest uncertainty – to hydrological models.

In this study, we investigate the Selenga River runoff in the XXI century. Therefore, the aim of this study is (a) to project and analyse the hydrological response to climate projections in the XXI century using an ensemble of GCMs and RCP-scenarios for the Selenga River basin; (b) to quantify the main sources of uncertainty of the obtained assessments of the possible runoff changes of Selenga River basin.

2 Case study

The Selenga River is the main tributary of the Lake Baikal, contributing to 50 % of its surface water inflow (Chalov et al., 2015). The vast transboundary river basin (447 000 km^2) is located in the northern part of Mongolia (approximately 64 % of the basin) and in Southern Siberia (Fig. 1). Natural landscapes in these areas varies from dry steppe (upstream) to Siberian taiga (downstream). The elevation ranges from 600 to 3000 m a.s.l. The local climate is extremely continental. The winters are long, dry, and cold, the mean monthly temperature of January is 23.5 °C. The summers are short and relatively warm, the mean monthly temperature of July is 16 °C. The annual precipitation amount ranges between 300 and 400 mm over the territory. Most of precipitation falls in July and August (about 45 % of the annual precipitation). Therefore, Selenga river runoff varies mostly because of variations of summer precipitation (Berezhnykh et al., 2012).

3 Material and methods

3.1 Description of hydrological model

The ECOMAG (ECOlogical Model for Applied Geophysics) is a spatially-distributed process-based hydrological model, which describing snow accumulation and melt, soil freezing and thawing, water infiltration into unfrozen and frozen soil, evapotranspiration, the thermal and water regime of soil and the overland, subsurface and channel flow with a daily time step (Motovilov et al., 1999; Motovilov, 2016).

The model accounts for measurable watershed characteristics such as surface elevation, slope, aspect, land cover and land use, soil and vegetation properties. The parameters are spatially distributed by partitioning the watershed into subbasins called elementary basins. Parameterization of the subgrid processes is described in Motovilov (2016). The model is driven by time series of daily air temperature, air humidity, and precipitation intensity.

The ECOMAG model has been applied for the Selenga River basin to assess the future impact of the regional climate change on the runoff using reanalysis dataset EWEMBI as inputs. The more detailed description of model parameters determination, results of calibration and validation against the observed data can be found in Gelfan and Millionshchikova (2018).

3.2 The hydrological response to climate projections using an ensemble of GCMs and RCP-scenarios

To assess the impact of climate changes on runoff in the Selenga River basin throughout the XXI century, we used the output of the GCMs from the Inter-Sectoral Impact Model Intercomparison Project Phase 2 (ISIMIP2), which operated under the RCP-scenarios.

The four RCPs, namely RCP2.6, RCP4.5, RCP6, and RCP8.5, were considered in accordance with the radiation impact level RCP in the year 2100: 2.6, 4.5, 6.0, and 8.5 $W m^{-2}$, respectively (IPCC, 2013). Four GCMs were used: HadGEM2-ES, IPSL-CM5A-LR, MIROC-ESMCHEM, GFDL-ESM2M. The original GCMs data were interpolated to a $0.5° \times 0.5°$ grid and bias-corrected (to reanalysis dataset WATCH) on the continental scale. The detailed description of this procedure is given in Hempel et al. (2013).

The reliability of runoff changes projections in the XXI century is determined by the correct reproduction of the GCM-driven hydrological model of observed climate conditions and runoff. The output of the hydrological model was evaluated in terms of daily runoff data observed at the basin outlet gauging station Selenga – Kabansk (Fig. 1).

In Millionshchikova (2019) it was shown that, despite the bias correction of the GCMs data against WATCH, the bias-errors in hydrological output simulated on the basis of these data still remained. Based on four climate models at four RCP-scenarios of the future anthropogenic impacts, an ensemble of possible climate changes projections for the Selenga River basin in the XXI century was obtained with correction rates that have been determined separately for each GCM relative to the historical low-flow period (1996–2005). The correction was performed by minimizing the bias-error of GCM-driven calculations on the regional scale (Gelfan et al., 2015) compared with EWEMBI-driven calculations. The assessment of hydrological consequences under climate changes in the Selenga River basin was simulated using the constructed hydrological model, driven by corrected GCMs

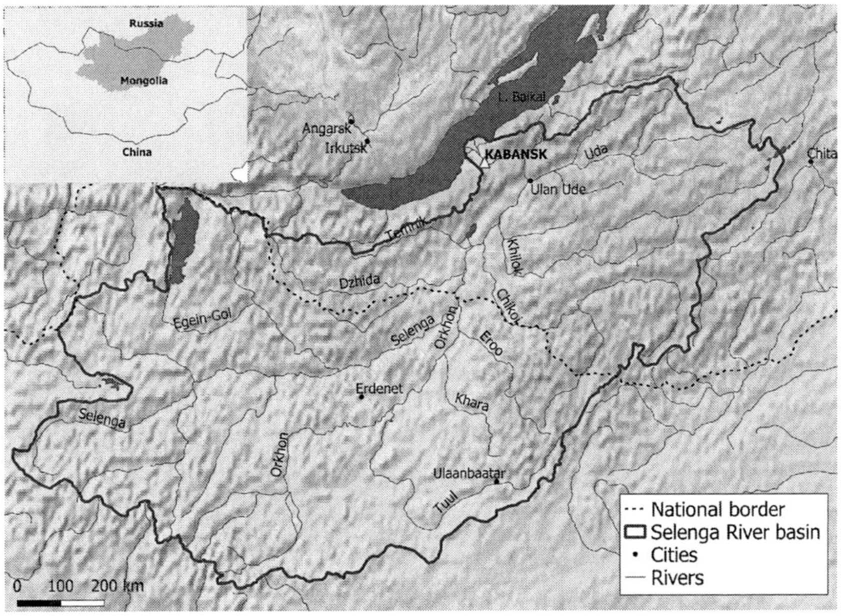

Figure 1. The Selenga River basin.

data, for three 30-year periods of the XXI century (2010–2039, 2040–2069 and 2070–2099). Selenga runoff anomalies, obtained by averaging runoff simulations under four GCM-based projections, were calculatated relative to the reference period (1996–2005). This allowed us to analyse differences in hydrological responses of the Selenga River basin to the specific behavior of the four RCP-scenarios.

3.3 Quantification of runoff projections uncertainty caused by climate projections

In this study we make use of ANOVA test given in Hattermann et al. (2018), to determine the sources of uncertainty and to express the total variation that can be attributed to the various factors. One can use different responses (in our case mean annual runoff anomaly) under climate changes to quantify sources of uncertainty and the significance of variation of the two factors: model uncertainty (differences between GCMs) and scenario uncertainty (variability of RCP-scenarios):

$$\mathrm{SST} = \frac{1}{N_{\mathrm{GCM}} N_{\mathrm{RCP}}} \sum_{i=1}^{N_{\mathrm{GCM}}} \sum_{j=1}^{N_{\mathrm{RCP}}} \left(Y_{ij} - \bar{Y}\right)^2 \tag{1}$$

where N_{GCM} – the number of GCMs; N_{RCP} – the number of RCP-scenarios; Y_{ij} – is the mean annual runoff anomaly value corresponding to the climate model i and RCP j; and $\bar{Y}$ – is the overall mean of runoff anomaly corresponding to all combinations of the climate models i and RCP-scenarios j.

SST can be further split into two main effects $\mathrm{SS}_{\mathrm{GCM}}$, $\mathrm{SS}_{\mathrm{RCP}}$, which are effects directly attributable to GCMs and RCP-scenarios, and into interaction term $\mathrm{SS}_{\mathrm{GCM*RCP}}$:

$$\mathrm{SST} = \mathrm{SS}_{\mathrm{GCM}} + \mathrm{SS}_{\mathrm{RCP}} + \mathrm{SS}_{\mathrm{GCM*RCP}}, \tag{2}$$

The main effects $\mathrm{SS}_{\mathrm{GCM}}$, $\mathrm{SS}_{\mathrm{RCP}}$ and interaction term $\mathrm{SS}_{\mathrm{GCM*RCP}}$ were calculated as:

$$\mathrm{SS}_{\mathrm{GCM}} = \frac{1}{N_{\mathrm{GCM}}} \sum_{i=1}^{N_{\mathrm{GCM}}} \left(\bar{Y}_{i0} - \bar{Y}\right)^2, \tag{3}$$

$$\mathrm{SS}_{\mathrm{RCP}} = \frac{1}{N_{\mathrm{RCP}}} \sum_{j=1}^{N_{\mathrm{RCP}}} \left(\bar{Y}_{0j} - \bar{Y}\right)^2, \tag{4}$$

$$\mathrm{SS}_{\mathrm{RCP*GCM}} = \frac{1}{N_{\mathrm{GCM}} N_{\mathrm{RCP}}} \sum_{i=1}^{N_{\mathrm{GCM}}} \sum_{j=1}^{N_{\mathrm{RCP}}} \left(Y_{ij} - \bar{Y}_{i0} - \bar{Y}_{0j} + \bar{Y}\right)^2. \tag{5}$$

where $\bar{Y}_{i0}$ – mean annual runoff anomaly when averaging over an ensemble of global climate models i under a fixed RCP-scenario; $\bar{Y}_{0j}$ – mean annual runoff anomaly when averaging over an ensemble of RCP-scenarios j under a fixed GCM.

4 Results

4.1 Selenga river runoff projections in the XXI century

We started with pre-testing an ability of the GCM-driven regional hydrological model to simulate historical runoff in

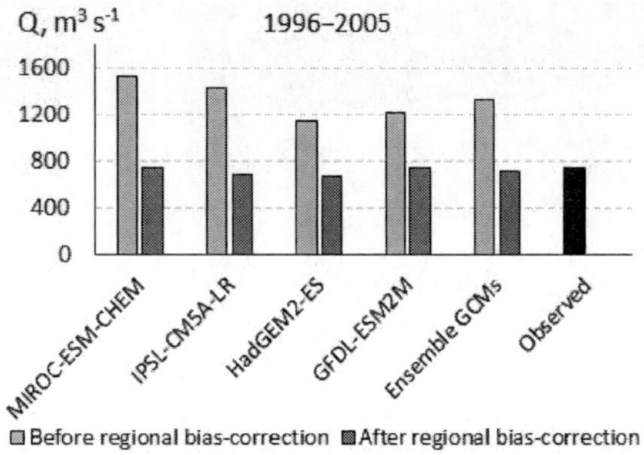

Figure 2. Results of evaluation ability of the GCM-driven regional hydrological model to simulate the observed mean annual runoff (Q, $\mathrm{m^3\,s^{-1}}$) in the Selenga R. at Kabansk gage.

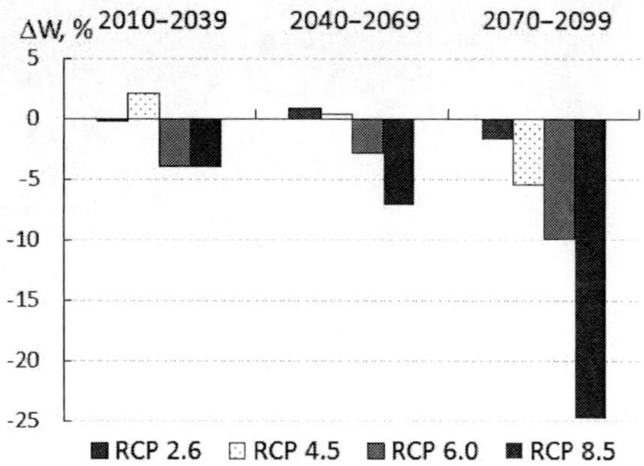

Figure 3. Mean annual runoff anomalies (ΔW, %) obtained relative to the period 1996–2005 under the RCP-scenarios.

the Selenga River basin. According to calculations, the observed and simulated runoff data do not match (Fig. 2). The observed mean annual Selenga runoff over 1996–2005 was 741 $\mathrm{m^3\,s^{-1}}$. The relative error of the river runoff calculation by data of global climate models was 80 % on average. After "regional" bias correction, the relative error of river runoff calculation by corrected data of GCMs ensemble has significantly decreased and equaled to 4 % on average.

Possible changes in the climate and runoff in the XXI century under different scenarios of greenhouse gas emissions were simulated as follows. To explain the regional runoff trends for the Selenga River basin, the ensemble of climate GCMs-RCP-based projections throughout the XXI century was obtained. The ensemble-averaged trends under different RCP-scenarios were used to estimate the mean annual anomalies of air temperature (°C) and precipitation (%) compared to the reference period (1996–2005). The estimated trends in the mean annual air temperature anomaly equals +1 °C per 100 years for RCP 2.6, +3 °C per 100 years for RCP 4.5, +4.2 °C per 100 years for RCP 6.0, and +7.3 °C per 100 years for RCP8.5. The estimated trends in the anomaly of the mean annual precipitation equals +2.5 % per 100 years for RCP 2.6, +8 % per 100 years for RCP 4.5, +9 % per 100 years for RCP 6.0, and +11 % per 100 years for RCP 8.5.

According to these climate projections, we simulated the possible runoff changes in the Selenga River basin. Selenga runoff anomalies are negative under all RCP scenarios and for all 30-year periods (Fig. 3). The several positive anomalies can be explained by a more significant increase of the projected evaporation simulated under the corresponding scenario. Overall, the runoff decrease will be more pronounced by the end of the XXI century. For instance, under RCP 8.5 the runoff will be decreased by ~ 25 % relative to the historical period.

4.2 Sources of uncertainty

The projected total uncertainty in impacts (changes in runoff anomalies) for three 30-year periods until the end of the century compared to the reference period (1996–2005) is shown in Fig. 4. The results illustrate that the model uncertainty makes a significantly larger contribution into the total variance of estimated runoff anomalies than scenario uncertainty, and the latter begins to grow by the end of the XXI century. Over the period 2010–2039, the model uncertainty contribution into the total variance of estimated runoff anomalies is 58 %. Over the next 30-year period, (2040–2069) the model uncertainty contribution rises to 88 % and by the end of the century (2070–2099) decreases to almost 68 %. Over the 2010–2039 and 2040–2069 periods the scenario uncertainty contribution into the total variance of estimated runoff anomalies is small (2 %), but by the end of the century (2070–2099) extremely grows (18 %). Over the 2010–2039 period, the contribution of the uncertainty associated with the interaction of model and scenario sources significantly exceeds the contribution of the scenario source (39 %). Whereas, by the end of the century (2070–2099) the contribution of the uncertainty associated with the interaction of model and scenario sources turns out to be comparable with scenario source (11 %–14 %) due to a decrease in the contribution of the model source into the total variance of estimated Selenga runoff anomalies.

5 Conclusions

In this study, we use the regional hydrological model ECOMAG to project the possible hydrological consequences in Selenga River basin under climate change in the XXI century using data from GCMs under different scenarios of greenhouse gas emissions. The model correctly reproduces runoff over the historical period driving by the bias-corrected (to

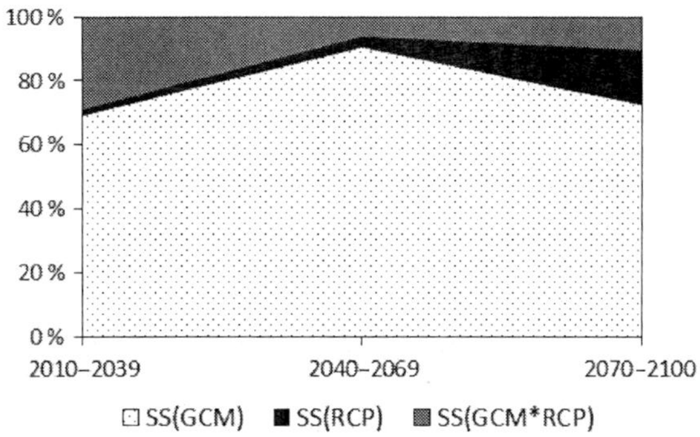

Figure 4. Sources of uncertainty in projected mean annual runoff anomalies for the Selenga River basin.

EWEMBI) climate data from selected GCMs. The obtained results provide a basis for the following assessment of hydrological model-based climate change impact for Selenga River basin. Depending on RCP-scenarios, an increase in air temperature and in precipitation equals 1–6 °C and 3 %–12 %, respectively. Overall, according to these climate projections, the low-flow period into Lake Baikal will be continued throughout the XXI century under all the RCP-scenarios.

However, uncertainty in the estimated Selenga river runoff anomalies in the XXI century should be accounted. ANOVA test was used to quantify sources of runoff projections uncertainty caused by climate projections. The results showed that the model uncertainty makes a significantly larger contribution to the total variance of estimated anomalies than scenario uncertainty. Moreover, scenario uncertainty contribution into the total variance increases by the end of the XXI century. This result is expected since the uncertainty of economic and technological global development forecasts increases by the end of the century.

Data availability. The datasets generated and analyzed during the current study are available from the corresponding author on reasonable request.

Competing interests. The author declares that there is no conflict of interest.

Special issue statement. This article is part of the special issue "Land use and climate change impacts on erosion and sediment transport". It is a result of the ICCE Symposium 2018 – Climate Change Impacts on Sediment Dynamics: Measurement, Modelling and Management, Moscow, Russia, 27–31 August 2018.

Financial support. This research has been supported by the Russian Fund for Basic Research (project 17-29-05027) and State scientific assignment 0147-2019-0001 (reg. AAAA-A18-118022090056-0).

References

Batima, P.: Climate Change Vulnerability and Adaptation in the Livestock Sector of Mongolia, Final Report, Project AS06, Assessments of Impacts and Adaptations to Climate Change (AIACC), International START Secretariat, Washington, DC, US, 85 pp., 2006.

Berezhnykh, T. V., Marchenko, O. Yu., Abasov, N. V., and Mordvinov, V. I.: Changes in the summertime atmospheric circulation over East Asia and formation of long-lasting low-water periods within the Selenga River Basin, Geogr. Nat. Resour., 33, 223–229, https://doi.org/10.1134/S1875372812030079, 2012.

Chalov, S. R., Jarsjo, J., Kasimov, N., Romanchenko A. O., Pietron, J., Thorslund J., and Promakhova, E. V.: Spatio-temporal variation of sediment transport in the Selenga River Basin, Mongolia and Russia, Environ. Earth Sci., 73, 663–680, https://doi.org/10.1007/s12665-014-3106-z, 2015.

Flato, G., Marotzke, J., Abiodun, B., Braconnot, P., Chou, S.C., Collins, W., Cox, P., Driouech, F., Emori, S., Eyring, V., Forest, C., Gleckler, P., Guilyardi, E., Jakob, C., Kattsov, V., Reason, C., and Rummukainen, M.: Evaluation of Climate Models, in: Climate Change 2013: The Physical Science Basis. Contribution of Working Group I to the Fifth Assessment Report of the Intergovernmental Panel on Climate Change, edited by: Stocker T. F., Qin D., Plattner, G.-K., Tignor, M., Allen, S. K., Boschung, J., Nauels, A., Xia, Y., Bex, V., and Midgley, P. M., Cambridge University Press, 741–882, 2013.

Frolova, N. L., Belyakova, P. A., Grigoriev, V. Y., Sazonov, A. A., Zotov, L. V., and Jarsjö, J.: Runoff fluctuations in the Selenga River Basin, Reg. Environ. Change., 17, 1965–1976, https://doi.org/10.1007/s10113-017-1199-0, 2017.

Gelfan, A., Semenov, V. A., Gusev, E., Motovilov, Y., Nasonova, O., Krylenko, I., and Kovalev, E.: Large-basin hydrological response to climate model outputs: uncertainty caused by internal atmospheric variability, Hydrol. Earth Syst. Sci., 19, 2737–2754, https://doi.org/10.5194/hess-19-2737-2015, 2015.

Gelfan, A., Gustafsson, D., Motovilov, Y., Arheimer, B., Kalugin, A., Krylenko, I., and Lavrenov, A.: Climate change impact on the water regime of two great Arctic rivers: modeling and uncertainty issues, Climatic Change, 141, 499–515, https://doi.org/10.1007/s10584-016-1710-5, 2017.

Gelfan, A. N. and Millionshchikova, T. D.: Validation of a Hydrological Model Intended for Impact Study: Problem Statement and Solution Example for Selenga River Basin, Water Resour., 45 (Sppl. 1), S90–S101, https://doi.org/10.1134/S0097807818050354, 2018.

Hattermann, F. F., Vetter, T., Breuer, L., Su Buda, Daggupati, P., Donnelly, C., Fekete, B., Flörke, F., Gosling, S. N., Hoffmann, P., Liersch, S., Masaki, Y., Motovilov, Y., Müller, C., Samaniego, L., Stacke, T., Wada, Y., Yang, T., and Krysnaova, V.: Sources of uncertainty in hydrological climate impact assessment: a cross-scale study, Environ. Res. Lett., 13, 015006, https://doi.org/10.1088/1748-9326/aa9938, 2018.

Hempel, S., Frieler, K., Warszawski, L., Schewe, J., and Piontek, F.: A trend-preserving bias correction – the ISI-MIP approach, Earth Syst. Dynam., 4, 219–236, https://doi.org/10.5194/esd-4-219-2013, 2013.

Hülsmann, L., Geyer, T., Schweitzer, C., Priess, J., and Karthe, D.: The effect of subarctic conditions on water resources: initial results and limitations of the SWAT model applied to the Kharaa River Basin in Northern Mongolia, Environ Earth Sci., 73, 581–592, https://doi.org/10.1007/s12665-014-3173-1, 2015.

IPCC: Climate Change 2013: The Physical Science Basis. Contribution of Working Group I to the Fifth Assessment Report of the Intergovernmental Panel on Climate Change, edited by: Stocker, T. F., Qin, D., Plattner, G. K., Tignor, M., Allen, S. K., Boschung, Y, Nauels, A., Xia, Y., Bex, V., and Midgley, P. M., Cambridge University Press, Cambridge, United Kingdom and New York, NY, USA, 1535 pp., 2013.

Joseph, J., Ghosh, S., Pathak, A., and Sahai, A. K.: Hydrologic impacts of climate change: Comparisons between hydrological parameter uncertainty and climate model uncertainty, J. Hydrol., 566, 1–22, https://doi.org/10.1016/j.jhydrol.2018.08.080, 2018.

Karthe, D., Chalov, S., Moreido, V., Pashkina, M., Romanchenko, A., Batbayar, G., Kalugin, A., Westphal, K., Malsy, M., and Flörke, M.: Assessment of runoff, water and sediment quality in the Selenga River basin aided by a web-based geoservice, Water Resour., 44, 399–416, https://doi.org/10.1134/S0097807817030113, 2017.

Krysanova, V., Donnelly, C., Gelfan, A., Gerten, D., Arheimer, B., Hattermann, F., and Kundzewicz, Z. W.: How the performance of hydrological models relates to credibility of projections under climate change, Hydrolog. Sci. J., 63, 696–720, https://doi.org/10.1080/02626667.2018.1446214, 2018.

Malsy, M., aus der Beek, T., and Flörke, M.: Global change impacts on Mongolian water resources, in: Beiträge zum 44. Jahrestreffen des Arbeitskreises Hydrologie, edited by: Chifflard, P., Cyffka, B., Karthe, D., and Wetzel, K. F., Augsburg, Geographica Augustana, 95–98, 2013.

Menzel, L., Aus der Beek, T., Törnros, T., Wimmer, F., and Gomboo, D.: Hydrological impact of climate and landuse change – results from the MoMo project, Int. conf, Uncertainties in water resource management: causes, technologies and consequences, edited by: Basandorj, B. and Oyunbaatar, D., Jakarta, IHP Tech. documents in hydrol, 1, 13–18, 2008.

Millionshchikova, T. D.: Selenga River Runoff Projections in the XXI Century: ECOMAG-Based Simulation Results, in: Climate Change Impacts on Hydrological Processes and Sediment Dynamics: Measurement, Modelling and Management, edited by: Chalov, S., Golosov, V., Li R., and Tsyplenkov, A., Springer Proceedings in Earth and Environmental Sciences, Springer, Cham, 61–66, https://doi.org/10.1007/978-3-030-03646-1_12, 2019.

Moreido, V. M. and Kalugin, A. S.: Assessing possible changes in Selenga R. water regime in the XXI century based on a runoff formation model, Water Resour., 44, 390–398, https://doi.org/10.1134/S0097807817030149, 2017.

Motovilov, Y., Gottschalk, L., Engeland, K., and Rodhe, A.: Validation of a distributed hydrological model against spatial observations, Agr. Forest Meteorol., 98–99, 257–277, https://doi.org/10.1016/S0168-1923(99)00102-1, 1999.

Motovilov, Y. G.: Hydrological simulation of river basins at different spatial scales: 1. Generalization and averaging algorithms, Water Resour., 43, 429–437, https://doi.org/10.1134/S0097807816030118, 2016.

Törnqvist, R., Jarsjö, J., Pietroń, J., Bring, A., Rogberg, P., Asokan, S. M., and Destouni, G.: Evolution of the hydro-climate system in the Lake Baikal basin, J. Hydrol., 519, 1953–1962, https://doi.org/10.1016/j.jhydrol.2014.09.074, 2014.

Proc. IAHS, 381, 113–119, 2019
https://doi.org/10.5194/piahs-381-113-2019

Modelling future hydroclimatic effects on the *Coregonus migratorius* spawning migration in the Selenga River and Lake Baikal

Vsevolod Moreydo[1], **Tatiana Millionshchikova**[1], **and Sergey Chalov**[2]

[1]Water Problems Institute of the Russian Academy of Sciences, Moscow, 119333, Russia
[2]Faculty of Geography, Moscow State University, Moscow, 119991, Russia

Correspondence: Vsevolod Moreydo (vsevolod.moreydo@iwp.ru)

Published: 1 August 2019

Abstract. Regional climate change affects the flow conditions in river basins which can impact the health of aquatic ecosystems. Potential impacts of future climate scenarios on *Coregonus migratorius* spawning migration in the Selenga River were assessed. A regional process-based hydrological model was used to reproduce the historical trends in the annual flow and assess its future changes under several climate change scenarios. Annual flow projections were used to identify preferential river reaches for spawning activity of the Arctic cisco (*Coregonus migratorius*), based on the significant negative correlation of spawning activity with the Selenga River streamflow. The applied methodology shows that the projected decline in runoff of 10 % to 25 % in XXI century may result in shifting of the spawning locations further upstream of the Ulan-Ude city, a local "pollution hotspot".

1 Introduction

In natural ecosystems, environmental abiotic drivers may play a crucial role in the development and survival of living organisms (Poff et al., 1997; Poff and Zimmerman, 2010; Stagl and Hattermann, 2016). In riverine ecosystems, the reproduction, foraging, maturation and recruitment of aquatic species are dependent on flow and temperature regime, both of which are vulnerable to natural and anthropogenic impacts (Gillenwater et al., 2006; Goto et al., 2015; Stagl and Hattermann, 2016; Wang et al., 2019). The health of aquatic ecosystems is strongly affected by catchment-scale human activities, such as dam construction (Fan et al., 2015; Wang et al., 2019; Wu et al., 2019), irrigation (Bączyk et al., 2018; Nhiwatiwa et al., 2017) and changes in hydro-climatic regime (Goode et al., 2012; Johnson et al., 2009; Statham, 2012). The observed changes in river flow due to these impacts can be evaluated using indicators of alteration in a riverine system (Gao et al., 2009; Richter et al., 1996). However, a commonly applied index for a basic assessment of the alteration of an aquatic system is the average annual streamflow discharge (Bragg et al., 2005).

The most interacting with the river flow are migratory fish species, such as salmonids, spawning in fresh water. The basic water quantity requirements for salmonid fish includes: flows (at appropriate times of the year), water depths, and flow velocities for spawning fish (Hendry et al., 2003). A reduction in flow causes a reduction in water levels and velocities, which is turn decreases available habitat and reduces habitat diversity (Dewson et al., 2007). Low flow conditions can be also attributed to other negative impacts such as increased sediment deposition and decrease in water quality. Numerous studies have used experimental flumes in the laboratory to examine the critical swimming speeds and optimal longitudinal velocities on the microhabitat level, including how velocity and turbulence influence on the variation in swimming behaviour (see Hockley et al., 2014). Whereas almost no studies were attributed to the effects of velocity to fish migration patterns of migratory fishes.

In recent studies a common methodology for assessment of future impacts of climate change on river flow and subsequently on the ecological state of the catchment system is involving a hydrological model forced by expected climate variables to simulate future flows and other ecohydrological indicators (Döll and Zhang, 2010; Gelfan et al., 2017; Moreido and Kalugin, 2017; Papadaki et al., 2016; Schneider et al., 2013; Shrestha et al., 2014; Thompson et al., 2014). Several studies have previously addressed the impact of future climate change on conditions of aquatic environments, such as anadromous fish morphology (Andres et al., 2019) and spawning success (Fenkes et al., 2016).

The present study examines hydroclimatic effects due to climate change on the seasonal spawning migration of Arctic cisco (*Coregonus migratorius)* in the Selenga Selenga River, which is the largest tributary of Lake Baikal and provides nearly a half of the annual inflow of water to the lake. The observed climate change impacts on the annual flow and regional hydrology (Karthe et al., 2014, 2017; Kasimov et al., 2017; Frolova et al, 2017) have a high potential to impair aquatic ecosystem functions and spawning migration of *Coregonus migratorius*. The Selenga River remains one of the rivers in the world free of large flow alteration projects (e.g. dams and water abstractions) of the comparable area (Frolova et al., 2017; Törnqvist et al., 2014).

Our assessment involved the construction of a regional-scale hydrological model to reproduce the current flow conditions of the Selenga River, forcing it with the future scenarios of climate change and subsequently applying the obtained results to a regression-based model the spawning migration of *Coregonus migratorius*.

2 Case study site

The Selenga River is a large transboundary river originating in Mongolia and flows across the Republic of Buryatia (Russia) to lake Baikal where it forms a large delta (Fig. 1). The Selenga River is 1024 km in length and drains an area of 450 000 km^2. Total annual flow into the lake is 29 km^3 and the highest flows occur in August–September. The Selenga River provides spawning habitat for the *Coregonus migratorius* (see Sect. 3.1). For the last two decades, annual flow of the Selenga River has decreased significantly (Frolova et al., 2017). With its diverse catchment flow generation conditions, ranging from dry steppe to mountainous taiga, and occasional permafrost patterns, the fate of the Selenga River and its ecosystem health under the expected climate change (Karthe et al., 2017; Törnqvist et al., 2014) is of crucial interest.

3 Materials and methods

3.1 *Coregonus migratorius* migration behavior assessment in current conditions

The Arctic cisco (*Coregonus migratorius*) is a migratory species of the *Salmonidae* family foraging in Lake Baikal and spawning in the tributaries. The Selenga and Upper Angara rivers contain the highest populations of Arctic cisco. estimated population of 1.0–3.7 million followed by spawning populations in the Barguzin and Bolshaya rivers (population around 0.1–0.7 million) (Bazov and Bazova, 2015).

Annual spawning migration occurs between the end of August and the end of October. According to multi-year monitoring programs there is a strong negative relationship ($R = -0.77$) between river stage and spawning migration distance during the spawning period (Bazov and Bazova, 2015; Sokolov and Kalyagin, 2001; Voronov, 1993). Preferable spawning conditions for *Coregonus* species include gravel bed (Roseman et al., 2007), low channel slope and flow velocities between 0.8–2.8 m s^{-1} (Bazov and Bazova, 2016). There has been no evidence in the local habitat studies of *Coregonus migratorius* spawning on inundated floodplains during high flows. During the spawning period, water temperature can vary between +9 and +20 °C and there is no significant correlation between temperature and spawning activity (Bazov and Bazova, 2015). This is likely due to the tendency of *Coregonus migratorius* to avoid additional energy loss due to overcoming the increased stream velocity.

Coregonus migratorius typically migrate 20 to 450 km upstream from the delta to spawn (Bazov and Bazova, 2015). A major industrial centre Ulan-Ude (capital of Buryatia Republic) is located 153 km upstream from the delta. Industrial activity represents a pollution hot spot at this location (Karthe et al., 2017). Accordingly, spawning activity is most pronounced in a reach river 140 km from the delta with an average distance of 110 km (Bazov and Bazova, 2010). During periods of low flow spawning activity can occur over a range of 175 to 450 km from the delta with an average distance of 240 km. The analysis of the dependency between the spawning distance and the mean water stage at the Kabansk gage in September showed a clear negative correlation of −0.77 (Bazov and Bazova, 2015), implying that the higher the stage, spawning activity occurs closer to the delta.

Bazov and Bazova (2015) outline the two types of *Coregonus migratorius* spawning behaviour:

1. "High-stage type" with the majority of spawning occurring downstream of Ulan-Ude, with the average distance of 110 ± 2 km, characterized by high water stage at Kabansk gage −340 ± 12 cm;

2. "Low-stage type" with the average spawning distance of 240 ± 26 km and water stage of 244 ± 15 cm.

3.2 The Selenga river flow modelling

ECOMAG modelling software was used to model current and assess future flow conditions of the Selenga river. The ECOMAG (ECOlogical Model for Applied Geophysics) is a spatially-distributed process-based hydrological model that considers hydrological processes such as rainfall/snowfall canopy interception, snow accumulation and melt, soil freezing and thawing, water infiltration into unfrozen and frozen soil, evapotranspiration, thermal and water regime of soil, overland, subsurface, ground and river flow with a daily time

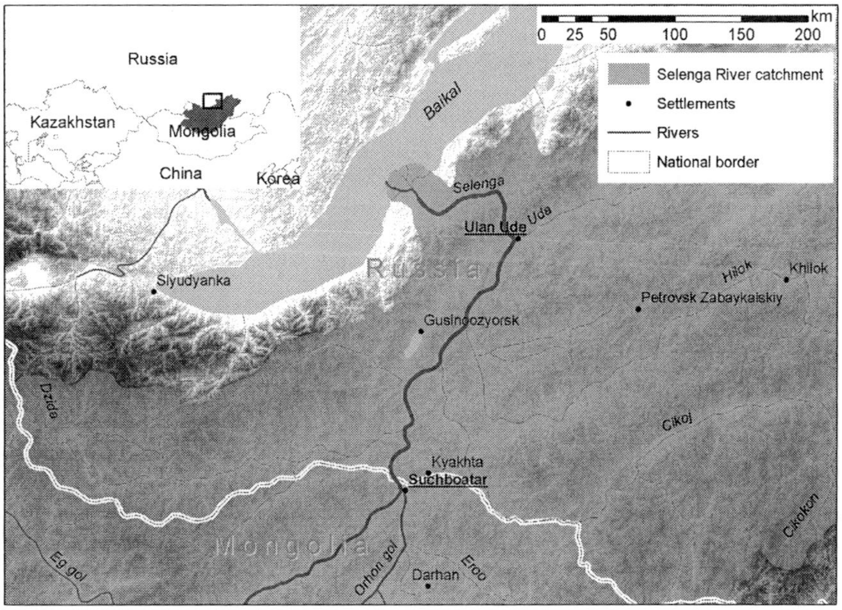

Figure 1. The location of the study region.

step (Motovilov et al., 1999; Motovilov, 2016). The model is capable of reproducing streamflow discharge at any point in the river network and is driven by time series data based on daily air temperature, humidity, and precipitation intensity. The ECOMAG model has been applied to the Selenga river basin to assess the future impact of the regional climate change on annual flow using EWEMBI reanalysis dataset (Lange, 2016) as input (Gelfan and Millionschikova, 2018).

The model parameters were calibrated against the observed daily streamflow discharge data at the most downstream hydrometric gauge located at Kabansk (Fig. 1) to reproduce the observed trends in annual flow of the Selenga River. The model showed good agreement between the measured and the calculated annual flow with the following efficiency estimates. For the calibration period (2000–2013) NSE = 0.82, $r = 0.93$ and BIAS = 0.22 %, for the validation period (1986–1999) NSE = 0.77, $r = 0.88$ and BIAS = 4.4 %.

3.3 Assessment of the Selenga flow and *Coregonus migratorius* spawning conditions under future climate changes

To assess annual flow dynamics in the Selenga river under future climate changes output from the General Circulation Models was used (GCMs) from the ISIMIP2: HadGEM2-ES, IPSL-CM5A-LR, MIROC-ESMCHEM, GFDL-ESM2M, which operated under different conditions of the expected external atmospheric radiation impacts (RCP-scenarios – Representative Concentration Pathways; IPCC, 2014). The original GCMs data were interpolated to a $0.5° \times 0.5°$ grid and bias-corrected on the continental scale (the bias-correction procedure concept is described in Hempel et al., 2013). The expected changes in the external parameters of the climate models – the scenarios of anthropogenic greenhouse gases' emissions – were introduced according to the radiation impact level for each of the four RCPs expected by 2100: RCP 2.6, RCP 4.5, RCP 6.0, RCP 8.5 (2.6, 4.5, 6.0, 8.5 W m^{-2}, respectively; IPCC, 2014). Based on the four GCMs forced by four RCPs, an ensemble of possible climate projections for the Selenga River basin in the XXI century was obtained with correction factors that have been determined for the historical low-flow period (1996–2005). The correction commonly consisted of minimizing the bias-error of GCM-based calculations on the regional scale (Gelfan et al., 2015) compared with the EWEMBI-based calculations.

This approach is unsuitable for intra-annual flow projections (Moreido and Kalugin, 2017) due to large uncertainties in the annual cycle described by the GCMs. Hence we used a linear regression model to change between the modelled annual flows and the September streamflow. The correlation coefficient between these variables is 0.91, which allows for such a transition. Next, the average streamflow value in September was converted into the mean stage value with the stage-discharge curve, established using long term data from the Kabansk gauging station. The analysis of the observed mean stage of the Selenga River in September and

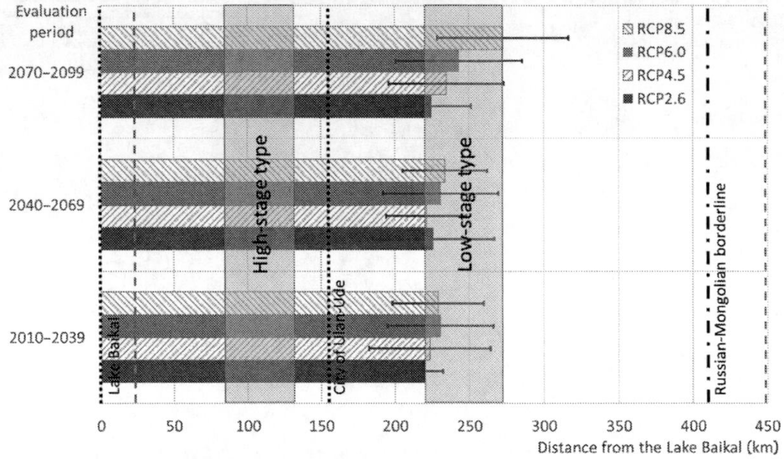

Figure 2. Projections of the Baikal omul spawning distance range in the XXI century. Horizontal bars show mean projected distance for different periods, whiskers show inter-model variance. Vertical grey dashed lines denote the historical minimum and maximum spawning ground location, the shaded areas show the two types of the migration distance from the assessment of 1984–2013 (see Sect. 3.1). Vertical black dotted line marks the location of the Ulan-Ude, black dash-dotted line shows the interstate borderline location.

the calculated using the described procedure showed a good agreement with correlation coefficient $r = 0.95$.

To predict *Coregonus migratorius* spawning features under future hydrological conditions we used a conceptual modelling approach. Accordingly, we employed a regression model (Bazov and Bazova, 2015) which links hydrological controls of the migration extent using relationship between distance of Baikal Coregonus migratorius spawning sites from the delta with the average water level of the Selenga river at most downstream gage in Kabansk in September.

4 Results and discussion

4.1 Streamflow modelling for XXI century

The future regional climate trends for the Selenga river based on the ensemble of four climate models' projections throughout the XXI century were obtained. The ensemble-averaged projections were used to assess the anomalies of air temperature (°C) and precipitation (%) over three thirty-year-long periods for the XXI century (2010–2039, 2040–2069, 2070–2099) as compared to the historical period of 1996–2005. Based on these results, the trend in the anomaly of the average annual air temperature in the Selenga Basin in the XXI century under different RCP-scenarios increased over time (Table 1). Accordingly, the trend in the anomaly of the average annual precipitation under different scenarios is projected to be positive (Table 1).

Using these projections as inputs to the hydrological model, we simulated flow changes in the Selenga River basin. According to all climate projections, mean annual flow

Table 1. Ensemble-averaged projections of air temperature (°C) and precipitation (%) anomalies for the XXI century in the Selenga river basin.

Variable	Relative concentration pathway (RCP)			
	2.6	4.5	6.0	8.5
T, °C	+1.0	+3.0	+4.2	+7.3
P, %	+2.5	+8.0	+9.0	+11.0

will decrease and this will be more pronounced in the second half of the XXI century. Under the RCP 8.5 scenario flow will decrease by 25 % relative to the historical period. It is assumed that the low-flow conditions into Lake Baikal will remain throughout the XXI century.

4.2 Assessment of the future alteration of the Baikal omul spawning migration

Using the obtained mean annual Selenga River streamflow discharge for future periods at the Kabansk gauge, we calculated the mean stage for these periods and subsequently – the possible alteration of the Baikal omul spawning distance using the methodology described in Sect. 3.1. Figure 2 shows the comparison of the current and the expected conditions of spawning grounds' location expressed as the distance upstream from the Selenga River delta.

The assessment of spawning locations based on the control catch data of 1984–2013 (Bazov and Bazova, 2015) showed that 25 % to 75 % of spawning grounds are located between 110 and 250 km from the delta. Our results show that according to the expected decrease in the Selenga River flow, the spawning migration distance will propagate further upstream. In the first period of 2010–2039 the distance assessed under different RCPs is projected at 219 to 230 km, in the second period – between 221 and 233 km, in the third – between 225 and 272 km. The inter-RCP and inter-model variance increase between the periods, with the most significant difference in the third period of 2070–2099 (Fig. 2).

On average, all the spawning grounds are projected to be located upstream of the Ulan-Ude city, which is ecologically favorable. Most of the projected distances fall into the "low-stage type" of the spawning behavior.

5 Conclusion

A modelling approach was used to assess the effect of climate change on streamflow and hydrological conditions for *Coregonus migratorius* spawning. Regional climate change projections from Global Circulation Models operating under different greenhouse gas forcing was determined and a regional hydrological model projecting future climate conditions was used with a regression-based spawning ground location model. The resulting projections show that annual flow of the Selenga River will decrease from 10 % to 25 % according to different climate change scenarios. These changes in river flow indicate that *Coregonus migratorius* will propagate further upstream of the Selenga River delta to more favorable spawning locations. Further investigations should include a detailed assessment of site scale physical, chemical and biological factors contributing to spawning success at the site, reach and watershed scale.

Data availability. The datasets generated and analyzed during the current study are available from the corresponding author on reasonable request.

Author contributions. VM and SC designed the study. TM performed the modelling and climate change assessment and wrote the corresponding section. VM performed the final assessment and wrote the rest of the paper. VM and SC made final approval for the publication.

Competing interests. The authors declare that they have no conflict of interest.

Special issue statement. This article is part of the special issue "Land use and climate change impacts on erosion and sediment transport". It is a result of the ICCE Symposium 2018 – Climate Change Impacts on Sediment Dynamics: Measurement, Modelling and Management, Moscow, Russia, 27–31 August 2018.

Financial support. This research has been supported by the Russian Fund for Basic Research (project 17-29-05027) and State scientific assignment #0147-2019-0003 (reg. #AAAA-A18-118022090105-5).

References

Andres, K. J., Chien, H., and Knouft, J. H.: Hydrology induces intraspecific variation in freshwater fish morphology under contemporary and future climate scenarios, Sci. Total Environ., 671, 421–430, https://doi.org/10.1016/j.scitotenv.2019.03.292, 2019.

Bączyk, A., Wagner, M., Okruszko, T., and Grygoruk, M.: Influence of technical maintenance measures on ecological status of agricultural lowland rivers – Systematic review and implications for river management, Sci. Total Environ., 627, 189–199, https://doi.org/10.1016/j.scitotenv.2018.01.235, 2018.

Bazov, A. V. and Bazova, N. V.: The monitoring assessment of the Baikal omul egg laying conditions in the Selenga river spawning grounds (lake Baikal basin), in: Biology, biotechnics of the breeding and resources conditions of the Coregonidae fish: Proceedings of the 7th International Applied Scientific Workshop, 16–18 February 2010, Tyumen, Russia, 70–74, Gosrybcentr, 2010 (in Russian).

Bazov, A. V. and Bazova, N. V.: Characteristics of spawning migration of the Baikal | Omul in the Selenga river according to long-term observations (1920–2014), Bull. Fish. Sci., 2, 18–28, 2015.

Bazov, A. V. and Bazova, N. V.: Selenga population of the Baikal omul: the past, the present and the future, Baikal Scientific Centre of the Siberian Branch of the Russian Academy of Sciences, Ulan-Ude, Russia, 2016.

Bragg, O. M., Black, A. R., Duck, R. W., and Rowan, J. S.: Approaching the physical-biological interface in rivers: a review of methods for ecological evaluation of flow regimes, Prog. Phys. Geogr. Earth Environ., 29, 506–531, https://doi.org/10.1191/0309133305pp460ra, 2005.

Dewson, Z. S., James, A. B. W., and Death, R. G.: A review of the consequences of decreased flow for instream habitat and macroinvertebrates, J. North Am. Benthol. Soc., 26, 401–415, https://doi.org/10.1899/06-110.1, 2007.

Döll, P. and Zhang, J.: Impact of climate change on freshwater ecosystems: a global-scale analysis of ecologically relevant river flow alterations, Hydrol. Earth Syst. Sci., 14, 783–799, https://doi.org/10.5194/hess-14-783-2010, 2010.

Fan, H., He, D., and Wang, H.: Environmental consequences of damming the mainstream Lancang-Mekong River: A review, Earth-Sci. Rev., 146, 77–91, https://doi.org/10.1016/J.EARSCIREV.2015.03.007, 2015.

Fenkes, M., Shiels, H. A., Fitzpatrick, J. L., and Nudds, R. L.: The potential impacts of migratory difficulty, including warmer waters and altered flow conditions, on the reproductive success of salmonid fishes, Comp. Biochem. Phys. A, 193, 11–21, https://doi.org/10.1016/j.cbpa.2015.11.012, 2016.

Frolova, N. L., Belyakova, P. A., Grigoriev, V. Y., Sazonov, A. A., Zotov, L. V., and Jarsjö, J.: Runoff fluctuations in the Selenga River Basin, Reg. Environ. Change, 17, 1965–1976, https://doi.org/10.1007/s10113-017-1199-0, 2017.

Gao, Y., Vogel, R. M., Kroll, C. N., Poff, N. L., and Olden, J. D.: Development of representative indicators of hydrologic alteration, J. Hydrol., 374, 136–147, https://doi.org/10.1016/j.jhydrol.2009.06.009, 2009.

Gelfan, A. N. and Millionschikova, T. D.: Validation of a Hydrological Model Intended for Impact Study: Problem Statement and Solution Example for Selenga River Basin, Suppl. 1, Water Resour., 45, S90–S101, 2018.

Gelfan, A. N., Semenov, V. A., Gusev, E., Motovilov, Y., Nasonova, O., Krylenko, I., and Kovalev, E.: Large-basin hydrological response to climate model outputs: uncertainty caused by internal atmospheric variability, Hydrol. Earth Syst. Sci., 19, 2737–2754, https://doi.org/10.5194/hess-19-2737-2015, 2015.

Gelfan, A. N., Gustafsson, D., Motovilov, Y., Arheimer, B., Kalugin, A., Krylenko, I., and Lavrenov, A.: Climate change impact on the water regime of two great Arctic rivers: modeling and uncertainty issues, Clim. Change, 141, 499–515, https://doi.org/10.1007/s10584-016-1710-5, 2017.

Gillenwater, D., Granata, T., and Zika, U.: GIS-based modeling of spawning habitat suitability for walleye in the Sandusky River, Ohio, and implications for dam removal and river restoration, Ecol. Eng., 28, 311–323, https://doi.org/10.1016/j.ecoleng.2006.08.003, 2006.

Goode, J. R., Luce, C. H., and Buffington, J. M.: Enhanced sediment delivery in a changing climate in semi-arid mountain basins: Implications for water resource management and aquatic habitat in the northern Rocky Mountains, Geomorphology, 139–140, 1–15, https://doi.org/10.1016/J.GEOMORPH.2011.06.021, 2012.

Goto, D., Hamel, M. J., Hammen, J. J., Rugg, M. L., Pegg, M. A., and Forbes, V. E.: Spatiotemporal variation in flow-dependent recruitment of long-lived riverine fish: Model development and evaluation, Ecol. Modell., 296, 79–92, https://doi.org/10.1016/j.ecolmodel.2014.10.026, 2015.

Hempel, S., Frieler, K., Warszawski, L., Schewe, J., and Piontek, F.: A trend-preserving bias correction – the ISI-MIP approach, Earth Syst. Dynam., 4, 219–236, https://doi.org/10.5194/esd-4-219-2013, 2013.

Hendry, K., Cragg-Hine, D., O'Grady, M., Sambrook, H., and Stephen, A.: Management of habitat for rehabilitation and enhancement of salmonid stocks, Fish. Res., 62, 171–192, https://doi.org/10.1016/S0165-7836(02)00161-3, 2003.

Hockley, F. A., Wilson, C. A. M. E., Brew, A., and Cable, J.: Fish responses to flow velocity and turbulence in relation to size, sex and parasite load, J. R. Soc. Interface, 11, 20130814, https://doi.org/10.1098/rsif.2013.0814, 2014.

IPCC: Climate Change 2013 – The Physical Science Basis, edited by: Stocker, T. F., Qin, D., Plattner, G.-K., Tignor, M., Allen, S. K., Boschung, J., Nauels, A., Xia, Y., Bex, V., and Midgley, P. M., Cambridge University Press, Cambridge, UK and New York, NY, USA, 1535 pp., 2014.

Johnson, A. C., Acreman, M. C., Dunbar, M. J., Feist, S. W., Giacomello, A. M., Gozlan, R. E., Hinsley, S. A., Ibbotson, A. T., Jarvie, H. P., Jones, J. I., Longshaw, M., Maberly, S. C., Marsh, T. J., Neal, C., Newman, J. R., Nunn, M. A., Pickup, R. W., Reynard, N. S., Sullivan, C. A., Sumpter, J. P., and Williams, R. J.: The British river of the future: How climate change and human activity might affect two contrasting river ecosystems in England, Sci. Total Environ., 407, 4787–4798, https://doi.org/10.1016/J.SCITOTENV.2009.05.018, 2009.

Karthe, D., Kasimov, N., Chalov, S., Shinkareva, G., Malsy, M., Menzel, L., Theuring, P., Hartwig, M., Schweitzer, C., Hofmann, J., Priess, J., and Lychagin, M.: Integrating Multi-Scale Data For The Assessment Of Water Availability And Quality In The Kharaa–Orkhon–Selenga River System, Geogr. Environ. Sustain., 3, 65–86, https://doi.org/10.24057/2071-9388-2014-7-3-40-49, 2014.

Karthe, D., Chalov, S., Moreido, V., Pashkina, M., Romanchenko, A., Batbayar, G., Kalugin, A., Westphal, K., Malsy, M., and Flörke, M.: Assessment of runoff, water and sediment quality in the Selenga River basin aided by a web-based geoservice, Water Resour., 44, 399–416, https://doi.org/10.1134/S0097807817030113, 2017.

Lange, S.: EartH2Observe, WFDEI and ERA-Interim data Merged and Bias-corrected for ISIMIP (EWEMBI), GFZ Data Serv., https://doi.org/10.5880/pik.2016.004, 2016.

Moreido, V. M. and Kalugin, A. S.: Assessing possible changes in Selenga R. water regime in the XXI century based on a runoff formation model, Water Resour., 44, 390–398, https://doi.org/10.1134/S0097807817030149, 2017.

Motovilov, Y. G.: Hydrological simulation of river basins at different spatial scales: 1. Generalization and averaging algorithms, Water Resour., 43, 429–437, https://doi.org/10.1134/S0097807816030118, 2016.

Motovilov, Y. G., Gottschalk, L., Engeland, K., and Rodhe, A.: Validation of a distributed hydrological model against spatial observations, Agr. Forest Meteorol., 98–99, 257–277, https://doi.org/10.1016/S0168-1923(99)00102-1, 1999.

Nhiwatiwa, T., Dalu, T., and Brendonck, L.: Impact of irrigation based sugarcane cultivation on the Chiredzi and Runde Rivers quality, Zimbabwe, Sci. Total Environ., 587–588, 316–325, https://doi.org/10.1016/j.scitotenv.2017.02.155, 2017.

Papadaki, C., Soulis, K., Muñoz-Mas, R., Martinez-Capel, F., Zogaris, S., Ntoanidis, L., and Dimitriou, E.: Potential impacts of climate change on flow regime and fish habitat in mountain rivers of the south-western Balkans, Sci. Total Environ., 540, 418–428, https://doi.org/10.1016/J.SCITOTENV.2015.06.134, 2016.

Poff, N. L. and Zimmerman, J. K. H.: Ecological responses to altered flow regimes: A literature review to inform the science and management of environmental flows, Freshw. Biol., 55, 194–205, https://doi.org/10.1111/j.1365-2427.2009.02272.x, 2010.

Poff, N. L., Allan, J. D., Bain, M. B., Karr, J. R., Prestegaard, K. L., Richter, B. D., Sparks, R. E., and Stromberg, J. C.: The Natural Flow Regime, Bioscience, 47, 769–784, https://doi.org/10.2307/1313099, 1997.

Richter, B. D., Baumgartner, J. V., Powell, J., and Braun, D. P.: A Method for Assessing Hydrologic Alteration within Ecosystems, Conserv. Biol., 10, 1163–1174, https://doi.org/10.1046/j.1523-1739.1996.10041163.x, 1996.

Roseman, E. F., Kennedy, G. W., Boase, J., Manny, B. A., Todd, T. N., and Stott, W.: Evidence of Lake Whitefish Spawning in the Detroit River: Implications for Habitat and Population Recovery, J. Great Lakes Res., 33, 397–406, https://doi.org/10.1016/S0380-1330(07)70064-3, 2007.

Schneider, C., Laizé, C. L. R., Acreman, M. C., and Flörke, M.: How will climate change modify river flow regimes in Europe?, Hydrol. Earth Syst. Sci., 17, 325–339, https://doi.org/10.5194/hess-17-325-2013, 2013.

Shrestha, R. R., Peters, D. L., and Schnorbus, M. A.: Evaluating the ability of a hydrologic model to replicate hydro-ecologically relevant indicators, Hydrol. Process., 28, 4294–4310, https://doi.org/10.1002/hyp.9997, 2014.

Sokolov, A. V. and Kalyagin, L. F.: General evaluation of the Baikal omul recruitment level, in State and problems of the artificial recruitment of the Baikal omul, IP Complex, Saint-Petersburg, Russia, 7–15, 2001.

Stagl, J. C. and Hattermann, F. F.: Impacts of Climate Change on Riverine Ecosystems: Alterations of Ecologically Relevant Flow Dynamics in the Danube River and Its Major Tributaries, Water (Switzerland), 8, 566, https://doi.org/10.3390/w8120566, 2016.

Statham, P. J.: Nutrients in estuaries – An overview and the potential impacts of climate change, Sci. Total Environ., 434, 213–227, https://doi.org/10.1016/J.SCITOTENV.2011.09.088, 2012.

Thompson, J. R., Laizé, C. L. R., Green, A. J., Acreman, M. C., and Kingston, D. G.: Climate change uncertainty in environmental flows for the Mekong River, Hydrol. Sci. J., 59, 935–954, https://doi.org/10.1080/02626667.2013.842074, 2014.

Törnqvist, R., Jarsjö, J., Pietroń, J., Bring, A., Rogberg, P., Asokan, S. M., and Destouni, G.: Evolution of the hydro-climate system in the Lake Baikal basin, J. Hydrol., 519, 1953–1962, https://doi.org/10.1016/j.jhydrol.2014.09.074, 2014.

Voronov, M. G.: Eco-biological basis for omul recruitment efficiency increase in the Selenga river in current conditions, Saint-Petersburg, Russia, 1993.

Wang, Y., Lei, X., Wen, X., Fang, G., Tan, Q., Tian, Y., Wang, C., and Wang, H.: Effects of damming and climatic change on the eco-hydrological system: A case study in the Yalong River, southwest China, Ecol. Indic., https://doi.org/10.1016/j.ecolind.2018.07.039, in press, 2019.

Wu, H., Chen, J., Xu, J., Zeng, G., Sang, L., Liu, Q., Yin, Z., Dai, J., Yin, D., Liang, J., and Ye, S.: Effects of dam construction on biodiversity: A review, J. Clean. Prod., 221, 480–489, https://doi.org/10.1016/j.jclepro.2019.03.001, 2019.

Proc. IAHS, 381, 121–126, 2019
https://doi.org/10.5194/piahs-381-121-2019

Time changes of dose equivalent rate above the soil surface as indication of natural attenuation processes

Alexei V. Konoplev[1], **Toshihiro Yoshihara**[2], **and Yoshifumi Wakiyama**[1]

[1]Institute of Environmental Radioactivity, Fukushima University, Fukushima, 960-1296, Japan
[2]Nuclear Risk Research Center, Central Research Institute of Electric Power Industry, Chiba, 270-1164, Japan

Correspondence: Alexei V. Konoplev (alexeikonoplev@gmail.com)

Published: 1 August 2019

Abstract. Eight sites in the Fukushima contaminated area were surveyed for long-term changes in D-shuttle dose equivalent rate above the soil surface during 2015–2017. D-shuttle readings in most cases decreased faster than if due to radioactive decay only. More rapid decrease can be explained by natural attenuation processes, such as erosion of the topsoil, the vertical migration of radionuclides in the soil and the deposition of cleaner sediments transported by surface runoff. According to the time dependencies of D-shuttle dose rate readings integral rate constants of the natural attenuation were estimated using the exponential trend-line. Estimated rate constants of natural attenuation ranged from 7.3×10^{-3} to $0.48\,yr^{-1}$, while the correspondent a half-dose rate decrease was 1.4–95 years.

1 Introduction

Today, 8 years after the accident at the Fukushima nuclear power plant, when emergency and short-term measures have been completed, the long-term dynamics of environmental radioactivity is becoming increasingly important. Currently, γ-radiation in areas contaminated in Fukushima is determined by two radioactive Cs isotopes – ^{137}Cs and ^{134}Cs released as a result of the accident (Malins et al., 2016; Konoplev et al., 2018; Okumura et al., 2016).

These radionuclides show similar behaviour in the environment since their physical and chemical properties, as well as speciation in depositions are the same.

The post-Chernobyl experience has shown that remediation of radioactively contaminated lands should be aimed at low cost, low intensity "passive" or low maintenance solutions rather than intrusive, and usually expensive, engineering techniques (IAEA, 2006a; Beresford et al., 2016). Monitored natural attenuation is one of such "passive" remediation options relying on natural processes that reduce the flux of radionuclides towards any given receptor (IAEA, 2006b; Konoplev et al., 2018). In this regard, processes of natural attenuation do not reduce the total amount of radionuclides in the environment. Rather they affect the distribution of radionuclides in space and time. Hydro-physical processes involved in natural attenuation (advection, diffusion, dispersion) may dilute radionuclides in the environment or redistribute them within the landscape (erosion and deposition), which, in turn, may lead to a gradual decrease of ambient dose equivalent rates (Malins et al., 2016; Konoplev et al., 2018).

A personal electronic cumulative dosimeter (D-shuttle; Chiyoda Technol. Corp., Japan) became available soon after the Fukushima accident in 2011. The D-shuttle dosimeter is able to measure γ-radiation with good accuracy at very fine resolution in the form of a dose equivalent rate ($> 0.01\,\mu Sv\,h^{-1}$) and storing data for more than 12 months (Čemusová et al., 2017).

Although the D-shuttle was developed to measure personal equivalent doses, most recently it was reported about the possibility of using the D-shuttle for measurement of environmental radiation (Konoplev et al., 2018; Kurita et al., 2018).

The purpose of this work is to study the dynamics of an ambient dose equivalent rates on the soil surface using D-shuttle dosimeters with a collimated lead shield and an estimate of the rates of natural attenuation due to the vertical and lateral migration of radiocesium.

2 Materials and methods

2.1 Study area

The locations of the study area and observation sites, which were described in detail in previous works (Konoplev et al., 2016, 2018), and map of air dose rate at 1 m height based on the results of the seventh airborne monitoring survey, 28 September 2013 (Secretariat of Nuclear Regulation Authority, 2013), are shown on Fig. 1.

The area contaminated after the accident at the Fukushima Dai-ichi Nuclear Power Plant (FDNPP) is characterized by a monsoon climate with annual precipitation ranging from 1100 to 1800 mm yr^{-1} during 2011–2016 according to the results of observations on five meteorological stations of the Japan Meteorological Agency – Haramachi, Iitate, Namie, Tsushima and Tomioka, which are located in the contaminated areas. The maximum precipitation falls on the typhoon season (mid-August–October) and the rainy season (end of May–mid-July). Temperatures are typical for a monsoon climate with a mild winter (the average monthly values of air temperatures are above zero) and with hot, rainy summer. There are no periods with soil freezing, and in conjunction with the large amounts of precipitation in summer and relatively high average annual air temperature, such conditions seem to contribute to the vertical migration of radiocesium in the soils (Konoplev et al., 2016).

It may be worth noting that the diversity of soil types in the Fukushima-contaminated areas is associated with a combination of rocks of different lithological composition, intense weathering and denudation due to high seismicity, as well as the steep slopes of mountain. Brown soils (under beech forest), ashy-volcanic, rich in humus, acidic allophonic (andosol) and leached brown soils are typical for the interfluve area. The valley bottoms are mainly used as paddy fields. Alluvial soils greatly modified due to years of land use are located here. Undisturbed alluvial soils are found in the levelled sections of the bottom of the river valley and along the canalized sections of the river beds, which is typical of intermountain depressions. Arable land, mainly rice fields, occupy about 12 % of the study area. They are mainly located in the vast depression and foothill lowlands.

2.2 Ambient dose equivalent rate measurements

Air dose equivalent rates at a height of 1 m were determined at all sites using an ALOKA pocket survey meter PDR-111. D-shuttle dosimeters with a collimated lead shield were used to measure and record hourly ambient gamma dose equivalent rates above soil surface at all selected sites. D-shuttle dosimeter has a size of 68 mm × 32 mm × 14 mm and weight of 23 g. These dosimeters employ semi-conductor technology using a Si-PIN diode (size of 12 mm × 14 mm) as the gamma-detector and are encased in a lead collimated shield (3 cm thick on the top, 3 cm thick sidewall and 2 cm thick underside) with a 3 cm × 3 cm collimation window in the underside shield opposite the diode. The devices were recording hourly dose equivalent rate inside the collimator corresponding to surface dose rate from a spot on the soil surface (with a diameter approximately equal to the installed device height) directly below the collimation window (Konoplev et al., 2018). D-shuttle dosimeters can operate up to one year without changing a battery. The dosimeters were installed at the heights of 10 to 60 cm above the soil surface using specially designed holders. D-shuttle dosimeters were installed on the sites under study from October 2015 to October 2017. Table 1 presents information about the observational sites and air dose equivalent rates at 1 m height on the sites under study at the time of installation of D-shuttle dosimeters. It should be noted that D-shuttle dose equivalent rate readings are essentially lower as compared with the air dose rate at 1 m height determined using ALOKA pocket survey meter PDR-111. However, D-shuttle readings are much more sensitive to changes in radiocesium inventory and its vertical distribution on a local spot on the soil surface covered by D-shuttle dosimeter with the collimated shield.

3 Results and discussion

The observed gamma dose equivalent rate can be attributed to ^{134}Cs and ^{137}Cs exclusively over the survey period 2015–2017. The initial ratio of isotopes $^{134}Cs/^{137}Cs$ in fallout immediately after the accident was equal 1 (Hirose, 2012). The temporary change in dose equivalent rate due to radioactive decay of Cs isotopes can be approximated by the equation below (Yoschenko et al., 2016; Konoplev et al., 2018):

$$\mathrm{DR}(t) = \mathrm{DR}(t_0) \times \frac{\left(e^{-\lambda_{137}\cdot t} + \alpha_{\frac{134}{137}} \cdot e^{-\lambda_{134}\cdot t}\right)}{\left(e^{-\lambda_{137}\cdot t_0} + \alpha_{\frac{134}{137}} \cdot e^{-\lambda_{134}\cdot t_0}\right)} \quad (1)$$

where t – current time after the accident; t_0 – time after the accident for the date of D-shuttle dosimeter installation; $\mathrm{DR}(t)$ – current dose rate, $\mathrm{DR}(t_0)$ – dose rate for the date of dosimeter installation; λ_{137} and λ_{134} – rate constants of radioactive decay for ^{137}Cs and ^{134}Cs, respectively; $\alpha_{134/137}$ – ratio of ^{134}Cs and ^{137}Cs gamma kerma equal 2.687 (Saito and Petoussi-Henss, 2014).

Figures 2 and 3 illustrate the dynamics of D-shuttle readings from soil surface at several sites under study together with data on precipitation and air temperature at the nearest meteorological station of the Japan Meteorological Agency (Namie, Tomioka or Haramachi depending on the site location). The dose rate in all cases, with the exception of in the floodplain of one of the Niida River (N2) in 2016 (Fig. 4), was decreased faster than if it was only due to radioactive decay.

A more rapid decrease in dose rate in the D-shuttle readings can be explained by natural attenuation processes, such as sheet and rill erosion, vertical migration of radionuclides

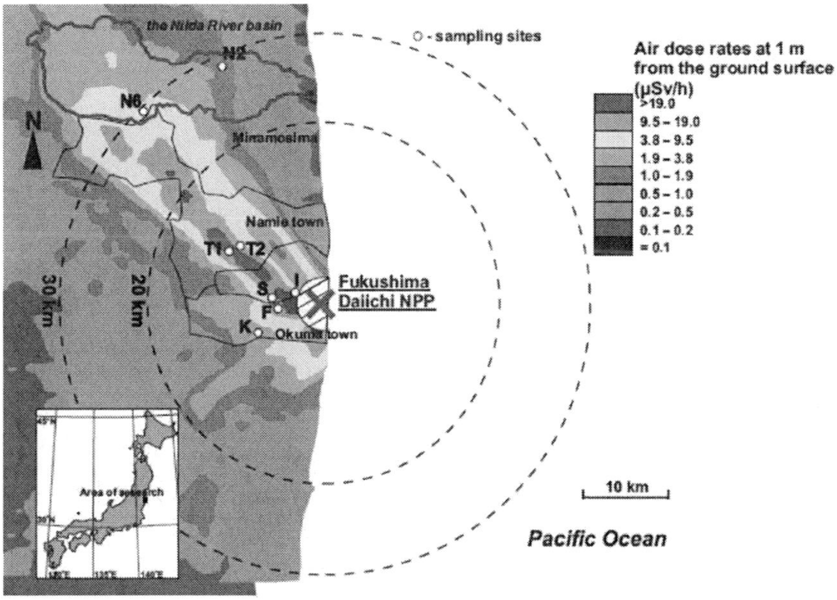

Figure 1. The map of the study area with locations of observation sites and air dose rates pattern at a distance 1 m from the surface which were determined based on the results of the seventh airborne monitoring survey, 28 September 2013 (Secretariat of Nuclear Regulation Authority, 2013). Red line delineates Niida River catchment. N2 and N6 are observation sites on the Niida River catchment; T1 and T2 are observation site on Takase River catchment; I is the site on the catchment of Inkyozaka pond; S is the site on the catchment of Suzuuchi pond; F is the site on the catchment of Funasawa pond; K is the site on the catchment of Kashiramori pond.

Table 1. Description of survey sites. The date format is month/year.

Location	Coordinates	Period of survey	Air dose rate[a], $\mu Sv\,h^{-1}$	^{137}Cs deposition, $kBq\,m^{-2}$	Soil type[b]
Catchment of Inkyozaka pond (I)	37.424800° N 141.017517° E	10/2016–09/2017	4.9	2100 ± 1000	Fluviosol
Catchment of Suzuuchi pond (S)	37.415767° N 140.979767° E	10/2016–07/2017	12.2	6400 ± 2200	Fluviosol
Catchment of Funasawa pond (F)	37.406050° N 140.986217° E	10/2016–10/2017	4.5	2900 ± 800	Terrestrial regosol
Catchment of Kashiramori pond (K)	37.379626° N 140.959180° E	10/2016–10/2017	0.9	900 ± 370	Andosol
Takase river catchment (T1)	37.465018° N 140.920988° E	10/2015–07/2016 12/2016–10/2017	18.2 17	5800 ± 400	Andosol
Takase river catchment (T2)	37.471189° N 140.935467° E	10/2015–07/2016 12/2016–10/2017	12 10	4400 ± 200	Terrestrial regosol
Niida river floodplain (N2)	37.660908° N 140.911855° E	02/2016–07/2016 12/2016–10/2017	1.0 0.4	280	Fluviosol
Hiso river floodplain (N6)	37.613650° N 140.801197° E	02/2016–07/2016	0.9	1660	Fluviosol

[a] Air dose rate on the date of the D-shuttle dosimeter installation; [b] according to soil classification of Japan (Obara et al., 2011).

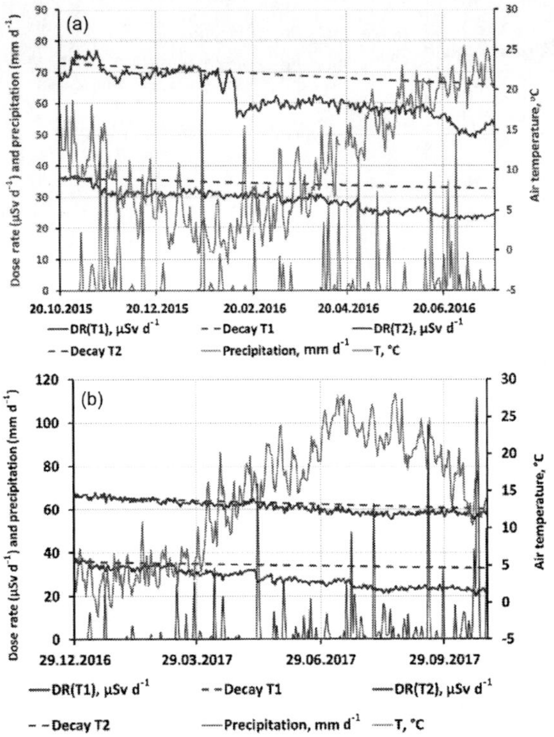

Figure 2. The time dependence of the D-shuttle readings (μSv d^{-1}) recorded with collimated shield from spot on the soil surface 10 cm below the dosimeter for the two sites T1 and T2 located in the Takase river catchment for two observation periods: October 2015 to July 2016 **(a)** and December 2016 to October 2017 **(b)** and data of daily precipitation (mm d^{-1}) and daily average air temperature (°C) recorded by Namie meteorological station of Japan Meteorological Agency for the similar time windows.

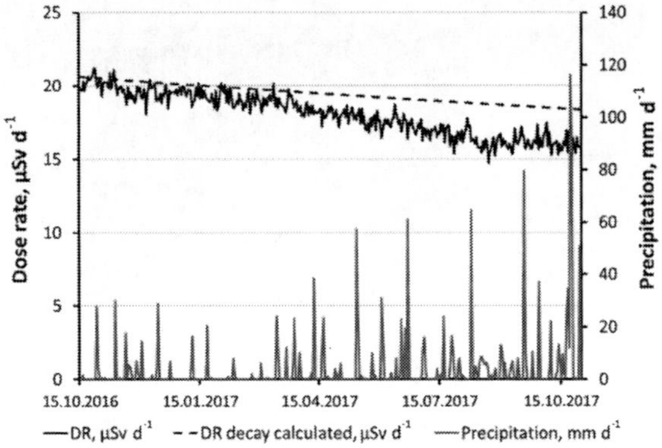

Figure 3. The time dependence of the D-shuttle readings (μSv d^{-1}) recorded from soil surface spot 10 cm below the dosimeter for the site located in the catchment area of Funasawa pond (F), Okuma town and data on daily precipitation (mm d^{-1}) recorded by Tomioka meteorological station of Japan Meteorological Agency for the similar time window.

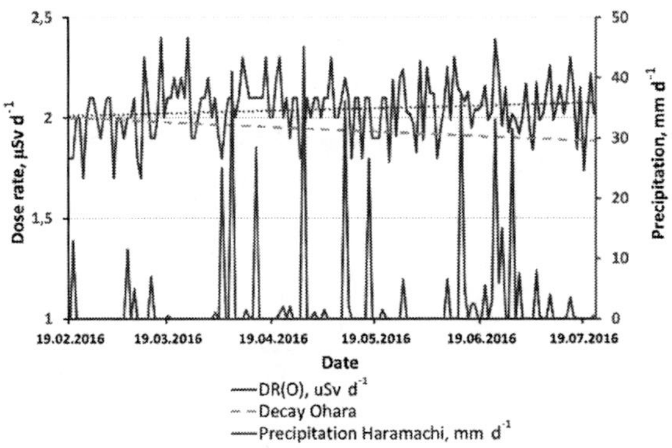

Figure 4. The time dependence of the D-shuttle readings (μSv d^{-1}) recorded from spot on the soil surface 20 cm below the dosimeter for the site, located on the floodplain of Niida river (N2) and data on daily precipitation (mm d^{-1}) for Haramachi meteorological station of Japan Meteorological Agency for the similar time window.

in the soil profile and deposition of cleaner sediments transported by surface runoff (IAEA, 2006b).

In addition to the time dependence of the D-shuttle readings, Figs. 2 and 3 demonstrate a hypothetical decrease in the correspondent dose equivalent rate over time solely due to the radioactive decay of ^{137}Cs and ^{134}Cs, which are not subject to any kind of migration.

In this case, the changes in dose equivalent rate are described by Eq. (1). From the time dependence in D-shuttle readings the integral rate constant of natural attenuation processes was estimated using an exponential trend line.

The calculated natural attenuation rate constants and its half-times are presented in Table 2. The estimated rate constants of natural attenuation processes ranged from 7.3×10^{-3} yr^{-1} on the catchment of the Suzuuchi pond in Okuma town to 0.48 yr^{-1} in the Takase River catchment with the correspondent half-reduction times being 1.4–95 years. The data shown in Fig. 2 for the catchment areas of the Takase River (sites T1 and T2) indicate a significant decrease in the dose equivalent rate during periods of flooding in November 2015 and from the end of May–June 2016. Also, a significant and sharp reduction in site T1 occurred during snowmelt and as a result of surface runoff in February 2016, when there was a sharp increase in air temperature. For the second period from December 2016 to October 2017 the natural attenuation processes at site T1 were slower, however they were faster at site T2. These variations seem to be explained by the fact that attenuation processes are strongly depend on hydrological characteristics and site-specific features of surface runoff.

There is only one site in floodplain of the Niida River (N2), where an increase in D-shuttle readings was observed due to

Table 2. Characteristics of γ dose equivalent rate attenuation for the sites under study. The date format is month/year.

Location	Period of survey	λ_{eff} [a], yr^{-1}	$T_{1/2}^{eff}$ [b], year	λ_{na} [c], yr^{-1}	$T_{1/2}^{na}$ [d], year
I	10/2016–09/2017	0.18	4.0	6.2×10^{-2}	11
S	10/2016–07/2017	0.12	5.7	7.3×10^{-3}	95
F	10/2016–10/2017	0.26	2.7	0.15	4.8
K	10/2016–10/2017	0.18	3.9	6.7×10^{-2}	10
T1	10/2015–07/2016	0.46	1.5	0.32	2.1
	12/2016–10/2017	0.18	3.8	7.5×10^{-2}	9.3
T2	10/2015–07/2016	0.52	1.3	0.38	1.8
	12/2016–10/2017	0.59	1.2	0.48	1.4
N2	12/2016–10/2017	0.13	5.3	2.1×10^{-2}	32
N6	02/2016–07/2016	0.35	2.0	0.21	3.3

[a] Effective rate constant of D-shuttle readings reduction; [b] half-life of effective reduction in D-shuttle readings; [c] rate constant of natural attenuation processes; [d] half-life of natural attenuation processes.

the deposition of contaminated sediments carried by the river from the upstream during the floods in the 2016 rainy season (Fig. 4).

4 Conclusions

D-shuttle dosimeters can be used to monitor long-term change in ambient gamma dose equivalent rates above the soil surface.

During the monitoring of eight sites in the Fukushima contaminated area the readings of the D-shuttle in most cases decreased faster than in the case of radioactive decay only. A more rapid reduction in dose equivalent rate is associated with natural attenuation processes such as sheet and rill erosion, vertical migration of radionuclides in the soil and deposition of cleaner sediments transported by surface runoff.

According to the time dependences in the D-shuttle readings, the integral rate constants of the natural attenuation were estimated using an exponential trend line. The estimated rate constants of natural attenuation ranged from $7.3 \times 10^{-3}\,yr^{-1}$ on a pond catchment in Okuma town to $0.48\,yr^{-1}$ in the Takase River catchment with the correspondent periods of dose rate half-reduction being 1.4–95 years. The data for the sites in the Takase River catchment indicate that a major reduction in dose rate was observed during flooding periods in November 2015 and from the end of May–June 2016, as well as during period of snowmelt. There is only one site on floodplain of the Niida River (N2) where an increase in D-shuttle dose rate readings were observed due to deposition of contaminated sediments carried by the river from upstream during the flood in the 2016 rainy season. Summing up, application of the D-shuttle for monitoring allows us to differentiate two distinct trends in the ambient dose equivalent rate, depending on specific lithological features of the monitoring spots, erosion and deposition of clean/contaminated sediments.

Data availability. The data can be accessed by request to authors.

Author contributions. AVK and TY designed the survey. AVK and YW carried out observations. AVK provided data analysis and prepared manuscript with contribution from all co-authors.

Competing interests. The authors declare that they have no conflict of interest.

Special issue statement. This article is part of the special issue "Land use and climate change impacts on erosion and sediment transport". It is a result of the ICCE Symposium 2018 – Climate Change Impacts on Sediment Dynamics: Measurement, Modelling and Management, Moscow, Russia, 27–31 August 2018.

Financial support. This research has been supported in part by a Grant-in-aid for Scientific Research from the Japan Society for the Promotion of Science (JSPS Projects 15H04621 and 18H03389).

References

Beresford, N., Fesenko, S., Konoplev, A., Skuterud, L., Smith, J. T., and Voigt, G.: Thirty years after the Chernobyl accident: what lessons have we learnt?, J. Environ. Radioactiv., 157, 77–89, https://doi.org/10.1016/j.jenvrad.2016.02.003, 2016.

Čemusová, Z., Ekendahl, D., and Judas, L.: Testing of the D-Shuttle personal dosimeter, Radiat. Meas., 106, 214–217, https://doi.org/10.1016/j.radmeas.2017.03.011, 2017.

Hirose, K.: 2011 Fukushima Dai-ichi nuclear power plant accident: summary of regional radioactive deposition monitoring results, J. Environ. Radioactiv., 111, 13–17, https://doi.org/10.1016/j.jenvrad.2011.09.003, 2012.

IAEA: Environmental consequences of the Chernobyl accident and their remediation: twenty years of experience. Report of the Chernobyl Forum Expert Group "Environment", IAEA, Vienna, Austria, 166 pp., 2006a.

IAEA: Applicability of Monitored Natural Attenuation at Radioactively Contaminated Sites. Technical Reports Series No. 445, IAEA, Vienna, Austria, 105 pp., 2006b.

Konoplev, A., Golosov, V., Laptev, G., Nanba, K., Onda, Y., Takase, T., Wakiyama, Y., and Yoshimura, K.: Behaviour of accidentally released radiocesium in soil-water environment: looking at Fukushima from a Chernobyl perspective, J. Environ. Radioactiv., 151, 568–578, https://doi.org/10.1016/j.jenvrad.2015.06.019, 2016.

Konoplev, A., Golosov, V., Wakiyama, Y., Takase, T., Yoschenko, V., Yoshihara, T., Parenyuk, O., Creswell, A., Ivanov, M., and Carradine, M.: Natural attenuation of

Fukushima-derived radiocesium in soils due to its vertical and lateral migration, J. Environ. Radioactiv., 186, 23–33, https://doi.org/10.1016/j.jenvrad.2017.06.019, 2018.

Kurita, K., Yamaguchi, M., Nagao, Y., Suzui, N., Yin, Y.-G., Yoshihara, T., and Kawachi, N.: Development of an Easy and Simple Method to Measure the Environmental Radioactivity in Trees with Efficient Personal Dosimeters, Radioisotopes, 67, 427–434, https://doi.org/10.3769/radioisotopes.67.427, 2018.

Malins, A., Kurikami H., Nakama, S., Saito, T., Konoplev, A., Golosov, V., Laptev, G., Nanba, K., Onda, Y., Takase, T., Wakiyama, Y., and Yoshimura, K.: Behaviour of accidentally released radiocesium in soil-water environment: looking at Fukushima from a Chernobyl perspective, J. Environ. Radioactiv., 151, 568–578, https://doi.org/10.1016/j.jenvrad.2015.06.019, 2016.

Obara, H., Ohkura, T., Takata, Y., Kohyama, K., Maejima, Y., and Hamazaki, T.: Comprehensive soil classification system of Japan first approximation, B. Natl. I. Agr. Sci. A, 29, 1–73, 2011 (in Japanese with English summary).

Okumura, M., Machida, M., and Kitamura, A.: Evaluation of ambient dose equivalent rates influenced by vertical and horizontal distribution of radioactive cesium in soil in Fukushima Prefecture, J. Environ. Radioactiv., 151, 38–49, https://doi.org/10.1016/j.jenvrad.2015.09.014, 2016.

Saito, K. and Petoussi-Henss, N.: Ambient dose equivalent conversation coefficients for radionuclides exponentially distributed in the ground, J. Nucl. Sci. Technol., 51, 1274–1287, https://doi.org/10.1080/00223131.2014.919885, 2014.

Secretariat of Nuclear Regulation Authority: Monitoring air dose rates from a series of aircraft surveys 30 months after the Fukushima Daiichi NPS accident. December 25, 2013, available at: http://radioactivity.nsr.go.jp/en/contents/9000/8700/view.html (last access: 21 June 2019), 2013.

Yoschenko, V., Nanba, K., Yoshida, S., Watanabe, Y., Takase, T., Sato, N., and Keitoku, K.: Morphological abnormalities in Japanese red pine (Pinus densiflora) at the territories contaminated as a result of the accident at Fukushima Dai-Ichi Nuclear Power Plant, J. Environ. Radioactiv., 165, 60–67, https://doi.org/10.1016/j.jenvrad.2016.09.006, 2016.

AUTHOR INDEX

International Association of Hydrological Sciences
Centre for Ecology and Hydrology
Wallingford, Oxfordshire OX10 8BB
United Kingdom

ISBN 978-1-5108-9544-7

CURRAN ASSOCIATES INC.
proceedings.com

European Distance and E-Learning Network Annual Conference (EDEN 2022)

Shaping the Digital Transformation of the Education Ecosystem in Europe

Tallinn, Estonia
20-22 June 2022

EUROPEAN DISTANCE AND E-LEARNING NETWORK